Energy Trading and Risk Management

Felix Müsgens · Alexander Bade

Energy Trading and Risk Management

Springer

Felix Müsgens
Brandenburg University of Technology
Cottbus-Senftenberg
Cottbus, Germany

Alexander Bade
Albstadt-Sigmaringen University
of Applied Sciences
Sigmaringen, Germany

ISBN 978-3-031-57237-1 ISBN 978-3-031-57238-8 (eBook)
https://doi.org/10.1007/978-3-031-57238-8

This Springer imprint is published by the registered company Springer Nature Switzerland AG
The registered company address is: Gewerbestrasse 11, 6330 Cham, Switzerland

Preface

This book focuses on trading, portfolio management and risk management for electricity and natural gas. While global trading volumes for both have been rising for decades, we think good textbooks on the subject are still lacking. At least in part, this can be explained by the properties of these commodities. They have exceptionally high storage costs, which sets them apart from other commodities, making cost-of-carry arbitrage almost impossible. Hence, the existing finance literature is of limited use to understand trading, pricing, and risk management on these markets. At the same time, high storage costs increase short-term price volatility and the need for a sound understanding of both pricing and risk management in these markets.

Our book fills this gap. After giving a brief introduction to the topic in Chap. 1, Chap. 2 is on market places for energy. It explains the unique properties of these markets, in particular the different time dimensions and their importance for trading and pricing. Both short- and long-term markets and their specific properties will be explained. The next chapter (Chap. 3) is on portfolio management as the core of energy trading on liberalised energy markets. We do this from different angles, starting with the simplest and ending with the portfolio management of a fully integrated energy supplier. Chapter 4 discusses risk management on energy markets. This chapter first defines the most relevant risks for energy trading and then focuses on the most important ones. Furthermore, the chapter explains the concept of risk capital, the steering and supervision of risks, and the processes needed to manage such risks professionally.

Both graduate students with knowledge in energy economics as well as practitioners new to energy trading will benefit from reading this book. It is based on practical experience the authors have gathered from working in power trading, portfolio management and consulting, and several classes they have taught at different universities. The book is "hands-on", following the different business models in energy trading: starting with prop trading (buying and selling on the wholesale market only), moving on to typical B2C sales activities, e.g. performed by municipalities, and finally adding generation capacity, typically power plants, to the portfolio.

All figures shown in this book are our own illustrations, if not noted otherwise. To enable readers to apply what they have learnt directly, we have also developed exercises. Some of them are explicitly mentioned in the main text of this book, but we have developed additional exercises as well. The full set of exercises, including the corresponding sample solutions, is available on the following website: https://www.b-tu.de/en/fg-energiewirtschaft/etrm. Access is password protected, the password is: **8y*3tWp**.

Cottbus, Germany Felix Müsgens
Sigmaringen, Germany Alexander Bade

Contents

List of Figures

Chapter 1
Introduction

In this chapter, we start the book by giving an overview of the term portfolio management, as we will use it in this book. We show how generation assets, utilities, wholesale markets and retail markets are interconnected and how portfolio management is at the heart of all activities in liberalised energy markets. Additionally, the characteristics of energy commodities, merit order formation and price formation are briefly explained.

Chapter 2 gives an overview of products and markets for energy products. We introduce the different markets, their main characteristics and market platforms where they are operated. We explain the different time horizons of energy trading, and how the different products are derived from that. On top of that, we show important characteristics of these markets that are needed to trade successfully, such as liquidity.

Chapter 3 of the book studies portfolio management. The analysis starts with proprietary trading on wholesale markets (when the trader neither owns physical assets nor sells to retail customers). Afterwards, complexity is increased step by step, analysing portfolio management of an energy utility with retail business only, an energy utility with generation assets only, and finally, analysing the management of an integrated portfolio, managing generation assets, buying and selling on the wholesale market and also selling energy to retail customers. The chapter concludes with the definition of portfolio values and the measurement of performance both for the whole company as well as parts of the company.

Chapter 4 of the book introduces risks and shows the importance of risk management. We describe the most important types of risk for the energy industry first and then discuss three important risks in detail: price risk, credit risk and product liquidity risk. We finish this chapter and the book by explaining the risk management process.

F. Müsgens and A. Bade, *Energy Trading and Risk Management*,
https://doi.org/10.1007/978-3-031-57238-8_1

1.1 Overview of Portfolio Management

Portfolio management is at the heart of energy trading on liberalised energy markets. Figure 1.1 provides an overview of where portfolio management is embedded in an energy company and also shows the topics we will discuss in Chap. 3, of this book. Right in the centre, we find the portfolio management, which is an operative service in a company that interacts with the rest of the organisation. Sometimes portfolio management is organised as a unit or division within a company. Sometimes, the service is provided by a third company specialised in consulting on portfolio management in the energy sector.

The tasks of the portfolio management process can be summarised as follows: If the company owns generation assets—this includes conventional power plants, renewable energy plants and gas production facilities—the portfolio management decides when to generate electricity and then sell this electricity on the wholesale market.

Even without generation assets, portfolio management interacts with the wholesale market, where energy commodities (gas and electricity mostly) are bought and sold. Finally, portfolio management can act on the retail market, where the company may sell energy to end customers such as households, businesses and industrial customers. Selling energy to retail customers is typically done by a sales department, and portfolio management is a fundamental part of this process as it procures the energy and is typically involved in pricing of the sales contracts.

When a company is active in all three fields, i.e. generation, as well as both the wholesale and the retail market, we say it manages an "integrated portfolio". While this is the most interesting case, it is also the most complex. For didactical reasons, we will start the analysis with a subset of the process and will increase complexity step-by-step afterwards.

Section 3.1 introduces portfolio management which operates exclusively on the wholesale market. This is called "proprietary trading" (or prop trading in short). A company such as a bank, a commodity trading house and a hedge fund can trade

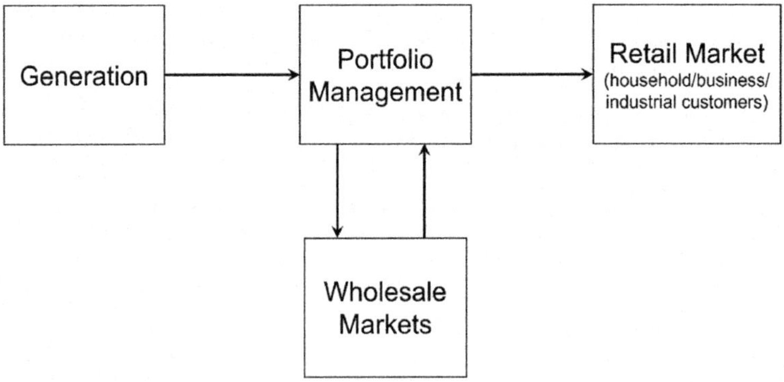

Fig. 1.1 Basic structure of the portfolio management interactions

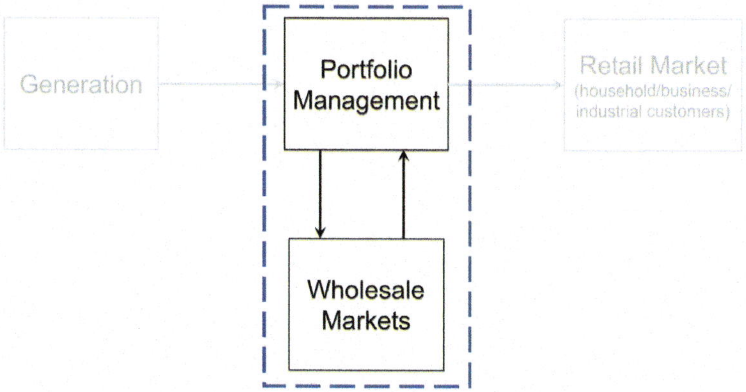

Fig. 1.2 Portfolio management solely with a wholesale market

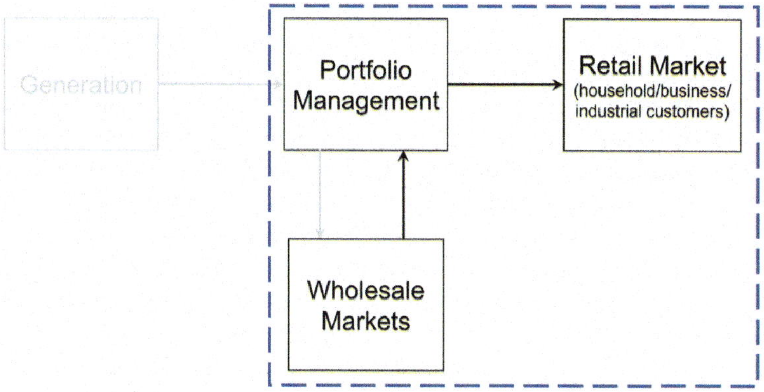

Fig. 1.3 Portfolio management with wholesale market and retail market

on the wholesale market at their own risk and benefit financially. In that case, the company's portfolio management service attempts to generate income by buying and selling energy contracts on the wholesale market. As such companies neither own any generation asset nor deliver energy to retail customers, they just trade contracts and close their positions before the delivery date. This is illustrated in Fig. 1.2.

Section 3.2 focuses on an energy utility with a retail focus and final consumers. This is illustrated in Fig. 1.3. We assume that the portfolio management of a utility buys energy on the wholesale market and the company sells this energy to retail customers.[1] This type of business is typical of municipal utilities that purchase both gas and electricity and sell them to end consumers, often in a specific geographical region.

[1] The company could also sell the energy first and procure it later. We will discuss this in the context of the "open position", which can be both positive and negative.

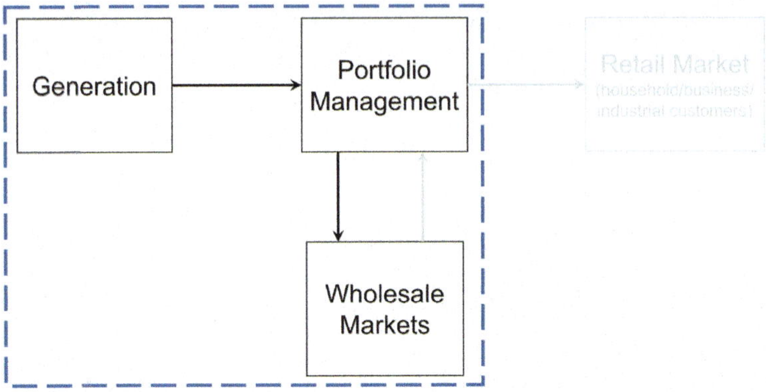

Fig. 1.4 Portfolio management with wholesale market and generation asset

In addition, similar portfolio management activities appear in companies consuming large amounts of energy, e.g. railway companies. Although they consume large amounts of energy, their main business model is to operate their trains, for example, and not to trade energy contracts on the wholesale market to generate revenue. Nonetheless, given the large sums of money involved, they often have a portfolio management team that is responsible for efficiently procuring the energy they need on the wholesale market. The same applies to large industrial and chemical companies that consume gigawatt-hours of energy per year.

Section 3.3 focuses on an energy utility with a generation asset. This is illustrated in Fig. 1.4. Here, we assume that a generation asset, either a conventional or a renewable power plant, produces electricity. In this case, the portfolio management is responsible for selling the generated energy on the wholesale market to maximise revenues. This process includes decisions on when and at which output level (full load or partial load) the plant operates over time. In a simplified way, this decision depends on whether the revenue is below or above production costs for a certain delivery period.

Finally, in Sect. 3.4, we will consider the integrated portfolio management process as illustrated in Fig. 1.1. This is typical for large utilities. They own generation assets, have retail customers, and have access to the wholesale market.

1.2 Characteristics of Energy Commodities

Some important characteristics set electricity and gas apart from other commodities. These characteristics need to be considered when trading them on markets:

- **Grid bound**: Electricity and gas need a grid to be transported from the production site to the consumer. In the case of electricity, generation must equal demand in real-time to guarantee a stable frequency in the electricity grid and prevent damage to electrical appliances. When it comes to gas, the grid provides a small buffer, but here too, injections and withdrawals must be precisely coordinated so that households are heated and industrial processes run.

- **Limited and costly storage**: It is possible to store electricity and gas (either directly or indirectly), but only in limited amounts, at significant cost and with efficiency losses.
- **Seasonality**: Electricity demand typically follows a seasonal profile. It is lower at night and higher during the day, and in so-called winter-peaking countries such as Germany, it is higher in winter and lower in summer. In summer-peaking countries, often in warm regions and with high electricity demand for cooling appliances, the load is typically highest on summer days. Renewable electricity generation exhibits seasonality on the supply side. For example, solar generation is high during the day and zero during the night—which thus is positively correlated with demand in summer-peaking countries. Usually, seasonality can be predicted with reasonable accuracy. In this book, we define seasonality as the estimable trend component of a time series (e.g. hourly demand).
- **Uncertainty**: Several factors affecting energy commodity prices are uncertain. For example, it is hard to predict with high accuracy if and to what extent the wind will blow a month from now, or how warm it will be next summer (and thus if the air conditioning systems will be widely used in a certain region, for instance). Forecasts are used to mitigate this effect. While they are increasingly accurate, there will always remain some error margin. In this book, we define uncertainty as the deviation of the value of a time series (e.g. hourly demand, which may include a seasonality component) from the expected value.

In addition, these factors have important consequences when designing markets, trading and price formation:

- **The geographical size of the market depends on grid characteristics and may vary over time**: Assume two areas (either countries or regions within a country) are interconnected by transmission lines, which allow the flow of energy from one area to the other. If the two areas are very well interconnected (i.e. there are sufficient transmission lines between them), a lot of energy can be traded between them. Whenever there is unused interconnector capacity between the areas, they tend to have the same electricity price and can be considered an integrated market area. On the other hand, if the two areas are less well connected (i.e. there are only a small number of transmission lines), electricity trading between them is limited. Whenever the available interconnector capacity provides a binding constraint to power flows between the areas, they will have different prices. Interestingly, whether two areas are an integrated market or not can change over time. For instance, if the internal prices in Germany and France do not differ a lot and there is still free transmission capacity, there will occur trade and prices in both areas will be equal at a certain hour. In other hours, the internal prices vary widely, and the capacity of the interconnection is too small to trade the required amount of energy to make the prices equal. In these hours, Germany and France exhibit different wholesale electricity prices.

- **Electricity (or natural gas) today is a different product compared to electricity (or natural gas) tomorrow**: The electricity that is consumed today is not the same product as the electricity that will be consumed tomorrow, next week or next month. Prices vary greatly, mostly due to seasonality and uncertainty. If electricity demand is low during the night, prices will also be low. In the case of natural gas, delivery in winter is a different product than delivery in summer. Typically—apart from the generation of electricity—, in summer natural gas is only used to heat water and for some industrial processes, but not for heating buildings and houses. This is a crucial finding, especially for readers with a background in finance, because most of the finance literature derives prices based on cost-of-carry arbitrage assumptions. Hence, the value of that literature is limited for non-storable commodities.

1.3 Merit Order and Price Formation

The merit order theory is fundamental to understanding how the price formation for energy commodities works. In Fig. 1.5 you can see a graphical example of a merit order, and you can see that the different technologies are sorted by their short-term generation cost in ascending order.

For each technology, both cost and capacity are relevant, because we need to make sure that supply matches demand at every moment. As we can see in the figure, less

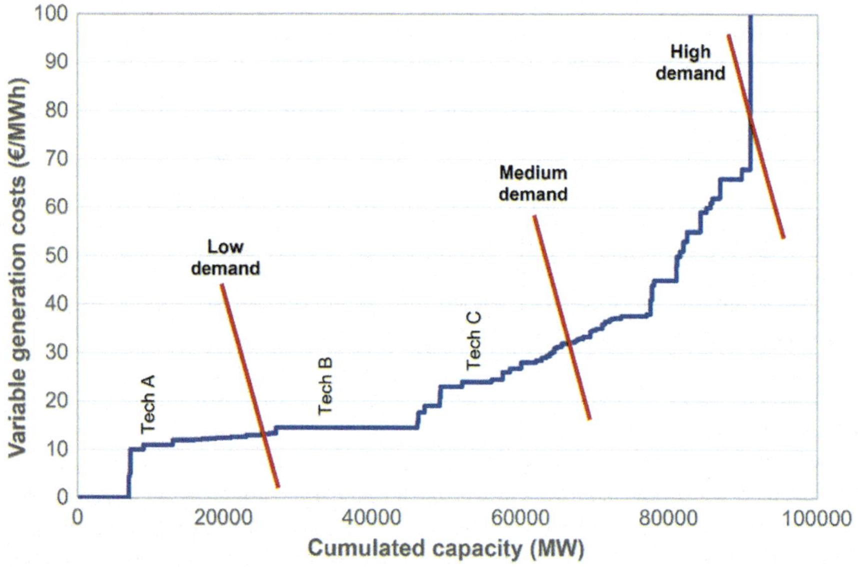

Fig. 1.5 Merit order formation example

capacity is required for low demand and therefore, the price is also low. When demand is high, more capacity is needed and the price increases.

This was only a brief introduction into the fundamental drivers of prices on the supply and demand side of electricity markets, but we want to emphasise that a solid understanding of the merit order theory and the price formation process in general is recommended for a better understanding of the following concepts. For any readers wanting to learn more, we recommend the textbooks "Economics of Power Systems-Fundamentals for Sustainable Energy" by Weber et al.,[2] "The Economics of Electricity Markets" by Biggar and Hesamzadeh[3] and "Power System Economics" by Stoft.[4]

References

Biggar, D. R., & Hesamzadeh, M. R. (2014). *The economics of electricity markets*. Wiley. https://doi.org/10.1002/9781118775745

Stoft, S. (2002). *Power system economics designing markets for electricity*. Wiley—IEEE Press.

Weber, C., Möst, D., & Fichtner, W. (2022). *Economics of power systems—Fundamentals for sustainable energy*. Springer.

[2] Weber et al. (2022).

[3] Biggar and Hesamzadeh (2014).

[4] Stoft (2002).

Chapter 2
Products and Markets

Energy commodities are traded on several different markets. These markets vary by time horizon between purchase and delivery (e.g. year-ahead, month-ahead or day-ahead), organisation of market (e.g. auction or continuous trading, power exchange or broker platform), traded commodity (e.g. electricity or gas), and type of financial contract (e.g. forward, future or option).

This chapter introduces the two vital time dimensions of energy trading, namely the delivery period and the trading period first. Regarding the delivery period, trading on day-ahead markets[1] takes place shortly before delivery and can be differentiated from delivery periods further in the future. In the day-ahead market, participants have significant knowledge about various factors on the supply side and the demand side as well as their own open position. Traded volumes on day-ahead markets are high in Europe, both for electricity and natural gas. Hence, we will start the description of markets with day-ahead markets, covering both electricity and natural gas. Then, we will discuss the longer time horizon and discusses continuous trading of long-term products, before differentiating different traded products. Lastly, we introduce the vital concept of liquidity and conclude the chapter with a brief discussion of the three key factors motivating energy trading.

[1] This market trades the 24 hours of the following day and is often referred to as a "spot market" for electricity. Note that strictly speaking, a spot market would entail immediate delivery of the product, i.e. on the spot. Hence, a day-ahead market is only an approximation of a spot market. In practice, there are also intraday markets, where trading takes place even shorter before delivery, but less volume is traded on these markets.

F. Müsgens and A. Bade, *Energy Trading and Risk Management*,
https://doi.org/10.1007/978-3-031-57238-8_2

2.1 Time Horizon of Energy Trading

When trading energy commodities, not one but two-time dimensions need to be differentiated: the trading period and the delivery period. The necessity to differentiate the two stems from the non-storability characteristic of energy as a commodity. This is one of the aspects setting energy commodities apart from most other traded commodities. The difference between trading period and delivery period is illustrated in Fig. 2.1.

It is important to distinguish between these two dimensions as they both lead to price fluctuations, but for different reasons. In the following, we will first take a closer look at the trading period and then discuss the delivery period, also known as the fulfilment period of a future product.

2.1.1 The Trading Period

The first-time dimension is the *trading period*, which is related to the time when a contract is concluded. For example, we can buy a front-year baseload contract today. We can also buy it tomorrow, next week or next month. Of course, we would still be buying the same product (front-year baseload electrical energy), but it has a different price depending on when we buy it.

Price differences over the trading period result from changes in the expectation regarding supply and demand factors determining the price: Today, the buyer of an electricity product has certain expectations for next year in terms of economic growth, investments in new power plants and other factors that can influence prices. However, these expectations may be different tomorrow (or even later today), next week, next month and any time before delivery starts. Therefore, the same product will trade at different prices during the trading period.

The concept of the trading period is similar to pricing in regular stock markets. Consider, for example, shares of companies such as Siemens or Google. Their prices may be different, depending on when they are bought. Stocks exhibit price variations, even though it is always a share of the same company—Siemens or Google—that is traded. These variations depend on the market expectation of the net present value of the underlying company and that expectation changes over time.

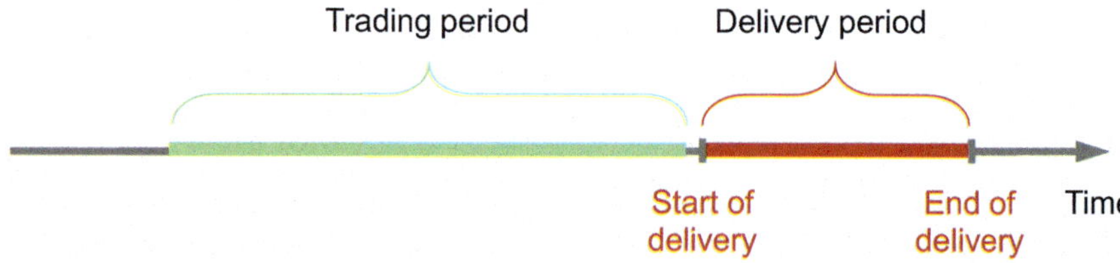

Fig. 2.1 Difference between trading and delivery period

Fig. 2.2 Price development of a base contract traded in the year before delivery

The products (e.g. a base year contract in the future) may be traded over a long time. Just to give an example, the front-year contract on the European Energy Exchange is quoted for up to ten years in the past. Obviously, the price can fluctuate significantly over such a long trading period. Historically, price fluctuations of more than 100% have been observed. The same year/product can be traded, for example, at 50 €/MWh and then a few months later at 100 €/MWh. The price always depends on the change in expectations for the delivery period. Again, this is comparable to a company's stock price which depends on the forecasts of the economic value of the company.

Figure 2.2 shows the exemplary variation in the price of a front-year contract. At the beginning of January, the contract opened with a settlement price shown in the red line. First it decreased a little and in February it increased again above the initial price in January. The closer the end of the year approached, the lower the price became. This contract finished significantly lower than in January and with significant fluctuations throughout the year. There is also some variation in the price within the day, and the lowest and highest daily price is also shown in the chart, as well as some moving averages for 200, 100 and 38 days.

2.1.2 *The Delivery Period*

The second time dimension is the *delivery period*, which answers the question of when the electricity is delivered (and consumed and produced, as electricity in a grid must be consumed at the time of production). This information allows us to classify the different products traded in the energy market at any given time. For instance, a front-year baseload contract specifies electrical energy delivered next year starting on the 1st of January and ending on the 31st of December. In the baseload contract, the energy is delivered 24/7, i.e. 24 hours on each day of the contract's delivery period. This is a very common product traded on almost all electricity markets.

The delivery period defines the product. You could say that electricity delivered during one month (e.g. January) is as different a product to electricity delivered during another month (e.g. July), just as Siemens shares are a different product to Google shares. Clearly, these two products will have different prices.

Since electricity can only be stored to a limited extent and at high cost, the prices in periods of high demand are ceteris paribus higher than in times of low demand. In other words, all things being equal, higher demand directly translates into higher prices. Especially in systems with considerable shares of renewable energy sources, supply also plays an important role in short-term price determination. Here are some examples for price differences of products in electricity markets:

- Electricity with delivery on weekends is cheaper than on working days, due to systematically lower demand.
- Due to the Christmas holidays, prices for December contracts are usually lower than delivery prices for January, again due to lower demand.
- Night hours are cheaper than day hours. This is because economic activity declines at night and low demand is again reflected in lower prices.
- Hours with a lot of renewable energy generation are ceteris paribus cheaper than hours with less renewable energy. Here, the reason is the additional supply from renewable energy sources at low variable costs.

2.2 Day-Ahead Markets for Electricity

Countries with a liberalised energy market, such as most European countries, typically trade electricity on day-ahead markets. Most often, these markets are technically organised as auctions. The auctions are performed for every hour of the following day, and they take place on power exchanges, such as, e.g. European Power Exchange (EPEX) or Nord Pool. On EPEX each day at 12:00[2] prices are determined for the 24 hours of the following day in 24 (mostly) independent auctions for different European regions or countries. EPEX covers several European markets, and they all have different auctions and different resulting prices.[3]

[2] For clarity purposes we will use a 24 h notation, i.e. 12:00 is 12:00 p.m.

[3] There is an ongoing process of European harmonization. As a part of this process, the daily day-ahead auctions are coupled in the so-called European hourly day-ahead coupled auction. More and more countries are participating in this coupled auction. For didactical reasons, we assume that day-ahead auctions will take place independently, i.e. prices will be determined by supply and demand in the respective region.

In preparation for each auction, potential buyers and sellers submit their bids,[4] indicating how much they are willing to buy or sell depending on the price. The exchange aggregates all bids and calculates the market price for every hour using a market clearing algorithm. In short, the market price is the point where the aggregated demand and supply curves intersect. Participation in these auctions is voluntary. The power exchange determines and publishes the different hourly prices for the next day soon after 12:00, which is why it is called the day-ahead market. Auction results are published shortly thereafter.

The day-ahead market has the following characteristics:

- The market is organised as an auction (a two-sided multi-unit uniform price auction).
- It is a physical market. If a company buys energy, it must specify in which transmission system operator's control area (country or region) it will be delivered.
- The smallest traded volume depends on the country and is typically between 0.1 and 1 MW.
- The exchange is the central counterparty for financial and physical settlement.
- The most liquidly traded products are individual hours as well as base and peak load.[5]
- Additionally, more complex block orders covering two or more hours are allowed. However, they are typically less frequently traded.

2.2.1 Aggregating Supply and Demand to a Market Price

An example of the whole process for an exemplary hour is shown in Fig. 2.3. Assume there are four different companies, numbered 1 to 4 in the figure, as market participants. Each company submits individual bids to buy or sell electricity. The table in the upper part of the figure shows these bids. Here, positive numbers represent buying bids, and negative numbers represent selling bids.

Looking at company 1 first, it wants to buy 200 MW for a duration of one hour[6] if the price is between zero and 9.9 €/MWh. In the range between 10 and 19.9 €/MWh, it still wants to buy, but only 100 MW. This can be rational because at that price range, it is either cheaper for the company to produce its own electricity (in case they own a generation unit), or to shift production to an hour in which the electricity prices are lower. If the price is exactly 20 €/MWh, company 1 does not want to buy or sell any

---

[4] Note that the use of the term "bid" depends on the context in this book. In some sections (e.g. here), we use it to comprise an entry to either buy or sell energy. In other sections, we differentiate between "bid" and "ask", where a bid comprises an entry to buy energy only, with the corresponding ask denoting the sell side.

[5] Base load covers supply in all 24 hours of a day, while peak load covers supply in the 12 hours between 08:00 and 20:00 of a day. This will be elaborated further in the following sections.

[6] Note that we analyse the situation in an exemplary hour. Hence, a demand of 200 MW for the duration of one hour is equal to 200 MWh of energy consumption during that hour. It could also be written as 200 MWh/h.

Price (€/MWh) / Company	0	9,9	10	19,9	20	20,1	20,2	149,9	150	3000
1	200,0	200,0	100,0	100,0	0,0	-80,0	-80,0	-80,0	-250,0	-250,0
2	150,0	150,0	50,0	50,0	5,0	5,0	0,0	0,0	-20,0	-20,0
3	-60,0	-60,0	-100,0	-100,0	-100,0	-175,0	-175,0	-175,0	-325,0	-325,0
4	200,0									200,0

Bids (MW)

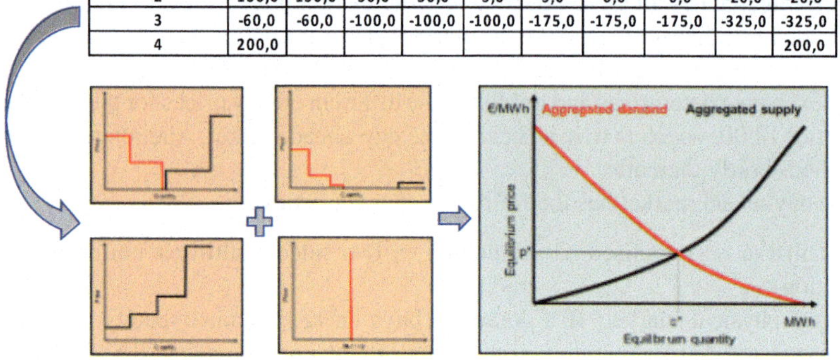

Fig. 2.3 Example of a bidding process and price establishment

electricity, the corresponding quantity bid is zero. At a price above 20 €/MWh, the company bids negative quantities, which means the company generates electricity and becomes a net seller on the wholesale market. At a price in the range between 20.01 and 149.9 €/MWh, they would like to sell 80 MW of electricity, and at prices above 150 €/MWh, they are interested in selling 250 MW.[7]

Similar reasoning applies to the second and third market participants in the example, i.e. company 2 and company 3. Company 4 wants to buy 200 MW of electricity regardless of the price. This means that it is advantageous for them if they can obtain energy at a favourable price, but that they are also prepared to pay the highest possible price to buy the energy if necessary. Again, the submission of bids is voluntary, and the bids of a market participant can vary from hour to hour: A company can be a consumer at some hours and become a seller at higher prices, both depending on the company's demand profile and the resulting prices.

In the next step, the exchange aggregates the bids, and the aggregated supply and demand curves are intersected to determine the market clearing (equilibrium) price and (equilibrium) quantities (lower part of Fig. 2.3). This process is performed on the daily day-ahead auction for each of the 24 hours of the following day. Let us assume in our example that there are more market participants than the four mentioned above and the resulting market clearing price[8] is 150 €/MWh.

Now we need to check what the different actors will do at the exemplary clearing price of 150 €/MWh. At this price, company 1 sells 250 MW, company 2 sells 20 MW, company 3 sells 325 MW and company 4 buys 200 MW. If these four market participants were the only ones in the market, 150 €/MWh could not be the equilibrium price because we would have a total of 595 MW sold but only

[7] Currently (October 2023), in the day-ahead auction at EPEX the minimum bidding price is −500 €/MWh and the maximum bidding price is 4000 €/MWh.

[8] Many exchanges and traders often just use the abbreviation MCP instead of "market clearing price".

200 MW bought. Thus, this price would be too high. We leave the determination of the market clearing price in the example with the four companies being the only market participants for an exercise.

In real-world energy auctions, there are usually more participants than in this example. We just assumed that there are other companies that are also willing to buy and sell energy at the equilibrium price in our example. To be precise, 395 more MW need to be bought than sold from these other market participants because the aggregated production must match the aggregated demand in equilibrium.

2.2.2 *Exemplary Day-Ahead Price Curves*

An example of the hourly result of a day-ahead auction is shown in Fig. 2.4 and Table 2.1. There, we can see the market prices for one day, split in 24 hours. During the night, prices are relatively low, in the example around 40 €/MWh. Then, in the early morning, prices start to rise to around 57 €/MWh (at 06:00) and to 80 €/MWh (at 07:00). The highest prices occur between 18:00 and 20:00, which is often the peak demand during winter months. In later hours, prices decrease again, and in the last two hours of the day, the price has already fallen below the daily average (roughly 68 €/MWh).

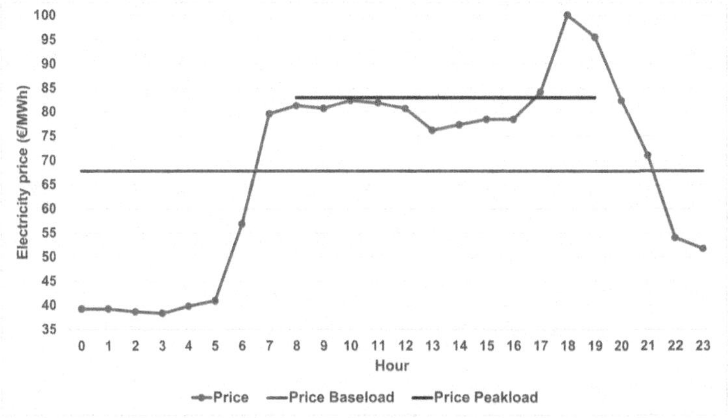

Fig. 2.4 Hourly prices in example day one

Table 2.1 Hourly prices in example day one

Hour	0	1	2	3	4	5	6	7	8	9	10	11
Price (€/MWh)	39	39	39	38	40	41	57	80	81	81	82	82
Hour	12	13	14	15	16	17	18	19	20	21	22	23
Price (€/MWh)	81	76	77	78	78	84	100	95	82	71	54	52

This daily average price of about 68 €/MWh in the example is represented by the blue line. A corresponding delivery during all hours of a designated time period (day, week, month or year) is defined as a base load delivery, or "base" for short. The price for this product is equal to the average of the whole traded period's hourly prices.

We can also see a red line, which is the so-called peak electricity price. Historically, the highest demands occur between 8:00 and 20:00, and as a result, also the highest prices. Therefore, this period was defined as peak. You can buy a peak product in the market, and it covers all hours from 8:00 to 20:00 on every working day of the year. Again, the price of this product equals the average of the hourly prices included in that time period.

Another exemplary day is shown in Fig. 2.5 and Table 2.2. The price structure looks similar in some respects, but different in others. Again, the day starts with relatively low prices during the night and then increases strongly in the early morning. In the seventh, eighth and ninth hours, prices are high, but in the following hours until 16:00 or 17:00, prices drop sharply to around 27 €/MWh. Afterwards, they rise again, and just as in exemplary day one, peak is reached in the evening, roughly at 20:00 or 21:00, with a price of more than 40 €/MWh, before prices decline again.

What is interesting about the second exemplary day is how much the hourly prices vary. As already mentioned, electricity today is a different product from electricity tomorrow, and—even more precisely—electricity at 3:00 differs from electricity

Fig. 2.5 Hourly prices in example day two

Table 2.2 Hourly prices in example day two

Hour	0	1	2	3	4	5	6	7	8	9	10	11
Price (€/MWh)	28	27	27	26	26	27	34	40	40	31	28	27
Hour	12	13	14	15	16	17	18	19	20	21	22	23
Price (€/MWh)	27	26	26	28	31	31	35	41	42	38	35	31

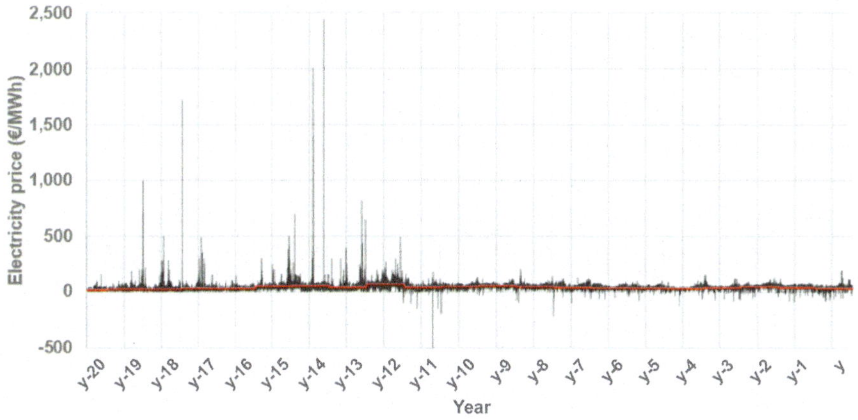

Fig. 2.6 Day-ahead electricity trading price development

at 20:00. In this example, the prices between the hour with the lowest price and the hour with the highest price differ by almost a factor of two. This is typical in practice and an important lesson to learn: The value of electricity can vary greatly between hours, and anyone interested in trading electricity should be aware of this fact. Consequently, it is necessary to specify precisely when the electricity must be delivered/produced/consumed.

The second relevant element about exemplary day two is what is driving the price decline during the day hours. This drop may be counterintuitive as we have already pointed out that the highest consumption typically occurs during the day. This question will be answered in an exercise.

Figure 2.6 shows the day-ahead electricity price development, for a time range of 20 years of a typical European market. Here you can see how strong the fluctuations are and how high the electricity prices can be at certain times. During most of this period, there was an upper limit for bids of 3000 €/MWh,[9] so the price could not rise above this limit, but in some hours the prices have almost reached it. However, most of the prices are below 100 €/MWh.

Another interesting fact is that electricity prices may be negative. In fact, electricity prices can change from positive to negative within hours. In this book, we refrain from discussing the energy economics behind this interesting fact. We rather focus on highlighting the importance for market participants that prices can become negative. If you want to consider negative prices in your calculations, you should be aware that many models do not take this into account, and just assume positive prices for every hour.

Figure 2.7 shows the annual average price for each year of the same sample period. They range between 20 and 50 €/MWh, except in $y + 8$ when they were higher, but this was an exceptional year (at least for the period shown).

[9] This limit was increased and is 4000 €/MWh as of May 2024 .

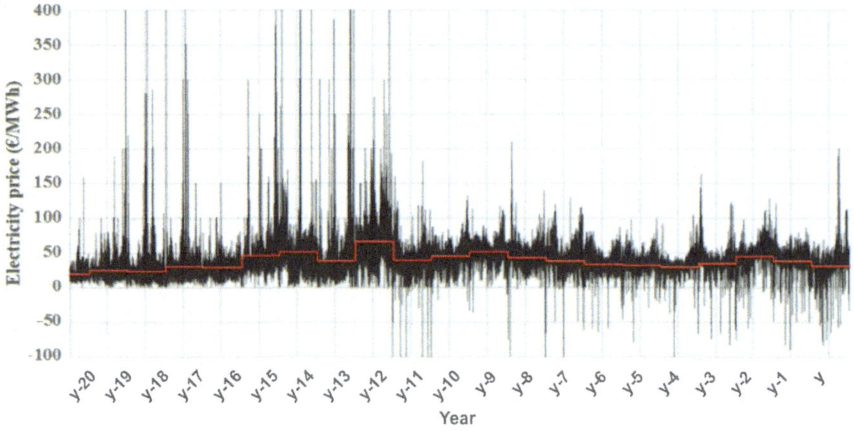

Fig. 2.7 Day-ahead electricity trading price annual average

These averages are driven by fundamental factors, such as investment costs. For instance, if the prices were persistently above the total costs for new installations, investors would build new generation assets. As a result, a sort of mean reversion process would take place.

It is unlikely that prices will be extremely low or extremely high for several years in a row because the market tends to adjust and that is what you can see in the graphs, particularly in Fig. 2.7. Prices tend to move in a certain range, but there are also a considerable number of price outliers that can be very high.

2.3 Day-Ahead Markets for Natural Gas

So far, we have mostly discussed electricity, but the major takeaways (particularly that supply and demand drive the price, that prices are influenced by seasonality and by uncertainty, etc.) are also valid for natural gas. Figure 2.8 shows an exemplary development of day-ahead gas trading prices. Natural gas prices exhibit less volatility than electricity prices because natural gas is easier and cheaper to store.

For the same reason, natural gas is mostly traded in the form of daily (and not hourly) products. Usually, a base delivery profile for a whole day is purchased, and natural gas is delivered for the 24 hours of the day. However, as shown in Fig. 2.9, there is some within-day pricing for balancing, provoking some volatility during the day. Natural gas is traded on various markets, such as the European Power Exchange, the Nord Pool Exchange and exchanges in the UK or the United States.

In conclusion, whenever a liberalised market is in place, the fundamental concepts previously (and to be) explained remain valid for natural gas.

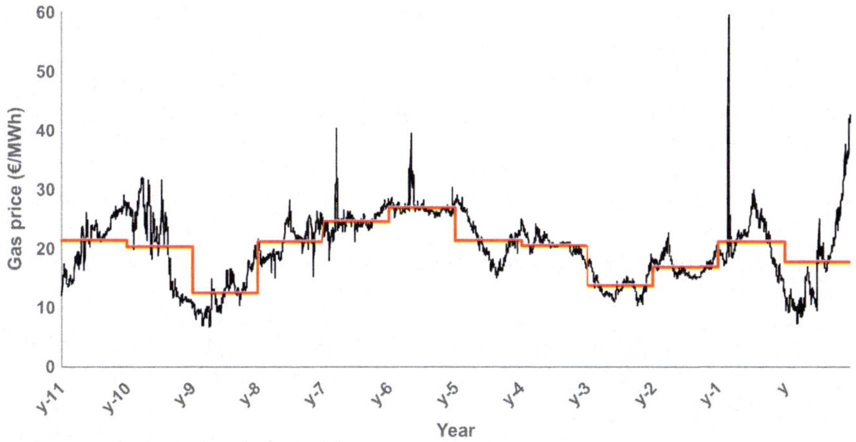

Fig. 2.8 Day-ahead gas trading price development

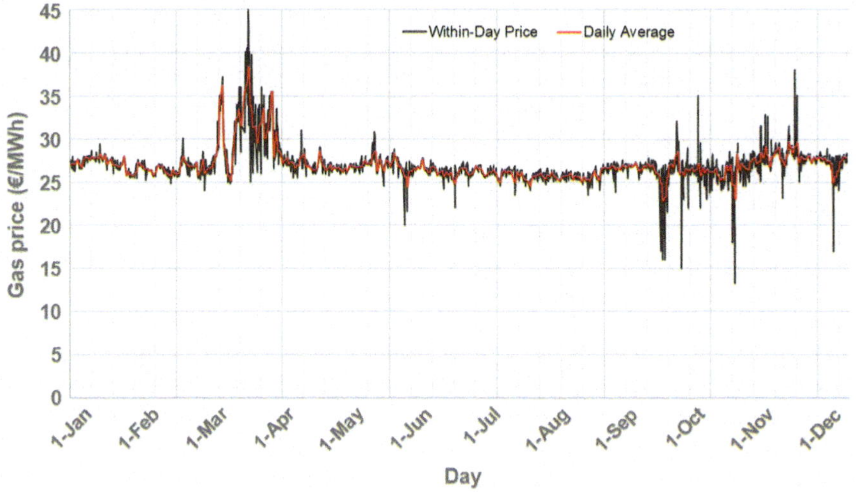

Fig. 2.9 Within-day gas trading prices of one example year

2.4 Continuous Trading on Long-Term Markets

In general, any trade in energy is forward-looking. This means that when a trade is executed, i.e. energy is bought or sold, delivery will take place sometime in the future. We have already discussed day-ahead markets in Sects. 2.2 and 2.3, organised as auctions. We will now analyse markets for products with a delivery date longer in the future. These markets are typically organised as continuous trading markets where market participants interact via voluntary bids and offers to buy and sell energy contracts.

This section will first explain how continuous trading is organised on platforms, then analyse the bid-ask spread and finally cover the pay-off profiles of long-term contracts.

2.4.1 Platforms for Continuous Trading

In Fig. 2.10 a typical platform is shown where long-term products are continuously traded. In the figure, it is possible to see how market participants buy and sell energy commodities on such platforms.

Basically, with this platform, the organisers of the market (a broker or a power exchange) provide "the cells on the display". In these cells, market participants who want to trade can submit bids and offers for a variety of products. For instance, in the left column block is Germany baseload, next is Germany peak load, Germany

Toggle Prod (all)	Germany BSLD						Germany PEAK			
	Qty	Bid	Ask	Qty	Last		Bid	Ask	Last	
Thu 08/10/y	25	127.45	128.00	25	127.45 ↔		168.05	168.35	168.05 ↔	
	25	127.45	129.05	25	126.75	1775	167.55	168.55	164.95	9000
	25	126.95	129.05	25	128.50	117.60	167.55	169.60	170.65	163.40
Fri 09/10/y	25	122.80	124.85	25	122.80 ↔		161.30	163.35	162.35 ↔	
Sat 10/10/y	25	77.50	81.15	25						
Sun 11/10/y	25	50.45	52.05	25	50.45 ↔		52.05			
Mon 12/10/y	25	89.50	129.05	25			106.15	162.35		
WkEnd 10/10-11/10	25	62.95	64.50	25	64.00 ↓					
WkEnd 17/10-18/10	25	62.45								
WkEnd 24/10-25/10										
Wk42 y	25	105.10	107.70	25	106.65 ↓		159.20	161.30	160.25 ↓	
Wk43 y	25	105.10	108.20	25	106.65 ↓		159.20	162.35	160.25 ↔	
Wk44 y	25	104.05	108.20	25	105.10 ↑		157.65	161.30		
Nov y	25	102.60	103.00	25	102.80 ↓		154.20	155.05	154.50	
	25	102.50	103.25	25	102.80	1455	154.00	155.45	152.95	550
	25	101.65	103.45	25	104.05	102.50	153.90	155.65	156.10	151.80
Dec y	15	94.15	94.70	15	94.15 ↔		133.50	133.80	133.70 ↓	
Jan y+1								150.35		
Feb y+1										
Mar y+1										
Apr y+1			89.50	5			118.65			
y+1 Q1	10	105.10	105.60	25	105.40 ↓		144.35	146.20	145.15 ↓	
	10	105.10	105.60	5	105.30	250	143.60		145.15 ↓	
	10	104.95	105.85	5	106.15	104.70	141.50		145.45	142.85
y+1 Q2	5	87.20	87.50	3	87.60 ↑		118.40	120.70		
y+1 Q3	5	92.60	94.15	10	93.65 ↑		132.65	134.75		
y+1 Q4	5	113.55	114.45	5	114.55 ↓		162.55	164.40		
y+2 Q1			125.10	10						
y+2 Q2	5	96.75								
y+2 Q3	5	101.05					146.40			
y+2 Q4	5	122.25	124.85	5						
y+1	5	100.00	100.10	5	100.00 ↔		140.70	141.20	141.50 ↔	
	5	99.90	100.20	5	100.00	669	140.70	141.30	141.10	55
	5	99.90	100.20	5	100.75	100.10	140.50	141.30	141.50	140.45
y+2	5	111.25	111.55	5	111.45 ↔		160.85	161.80	161.80 ↔	
	5	111.25	111.55	5	111.25	71	160.65	162.00	161.80	20
	5	111.15	111.65	5	111.75	53.45	160.65	162.35	161.80	77.00
y+3	5	118.40	119.05	5	57.10 ↑		170.15	171.20		

Toggle Prod (all)	Germany O/PEAK				Germany 0-6				Germany 20-24			
	Bid	Ask	Last		Bid	Ask	Last		Bid	Ask	Last	
Thu 08/10/y	86.90	89.50	87.95 ↑		54.65	55.35	55.35 ↑		112.40	112.90	112.90 ↑	
	86.90	91.55	87.41	425	54.65	55.65	54.10	175	112.40	113.40	112.35	350
	86.35		90.00	81.15	54.10	55.65	55.35	0.00	112.40	118.65	113.10	0.00
Fri 09/10/y	74.90	89.50			49.95	62.45			93.65			
Sat 10/10/y						60.35						
Sun 11/10/y					-20.80							
Mon 12/10/y		87.40				58.25						
WkEnd 10/10-11/10												
WkEnd 17/10-18/10												
WkEnd 24/10-25/10												
Wk42 y	74.40	91.55										
Wk43 y												
Wk44 y					API 2				TTF Hi Cal 51.6			
Nov y	74.90	75.95			152.45	153.15	152.45 ↔		25.40	25.60	25.49 ↔	
	74.90	77.00		0			152.45	10	25.30	25.70	25.49	150
	74.70	78.05		75.95			152.45	151.90	25.30	25.70	25.49	25.80
Dec y		72.85	71.80 ↓						28.10	28.60	28.30 ↓	
Jan y+1									29.75	30.90		
Feb y+1									29.65	31.20		
Mar y+1									28.95	30.40		
Apr y+1									28.10	30.20		
y+1 Q1	82.20	84.80			160.55	162.00	160.65 ↓		29.95	30.30	30.05 ↓	
				0	160.45	162.75	160.55	35	29.75	30.40	30.05	30
				80.10	159.85		160.75	160.35		30.40	30.05	30.30
y+1 Q2					169.10	170.35	169.80 ↓		28.60	30.30		
y+1 Q3					179.00	180.25	179.00 ↔		28.40	29.65		
y+1 Q4					186.25	188.15			38.90	40.60		
y+2 Q1					191.45	197.00			43.20	44.95		
y+2 Q2												
y+2 Q3												
y+2 Q4									43.90	46.00		
y+1		79.10	77.52 ↓		173.45	174.40	173.80		31.85	32.15		
			77.50	50	173.25	174.80	173.45	45	31.75	32.25		0
			77.75	76.50	173.25	174.80	174.30	173.90	31.65	32.25		31.65
y+2					205.50	206.65			42.25	42.55		
				0	205.10	208.10		0	42.15	105.20		0
				83.55	204.80	209.15		201.85	41.95	42.85		41.85
y+3					223.20	225.80			45.80	46.85		

Time	Instrument	Price	Aggressor Company	Qty
10:33:32	Germany BSLD y+1		[S]	5
10:33:32	Germany BSLD y+1		[S]	5
10:33:29	Germany BSLD y+1		[S]	5
10:33:14	Germany BSLD Dec-y		[S]	25
10:33:05	Germany BSLD Dec-y		[S]	10
10:32:50	E.ON GT H-GAS WD		[B]	150
10:31:56	Germany BSLD y+1		[S]	5
10:31:55	Germany PEAK y+1 Q1		[S]	10
10:31:50	Germany BSLD y+1		[S]	5

	EUA ECX			
Dec y	28.65	28.75	28.70 ↔	
	28.60	28.80	28.70	1290
	28.60	28.80	28.95	28.40
Dec y+1	29.20	29.30	29.30 ↔	
		29.40	29.30	725
			29.45	29.25
Dec y+2	30.45	30.55	30.51 ↓	
			30.51	175
			30.59	30.07

Fig. 2.10 Generic market platform

Toggle Prod (all)	Germany BSLD					
	Qty	Bid	Ask	Qty	Last	
y+1	5	100.00	100.10	5	100.00	↔
	5	99.90	100.20	5	100.00	669
	5	99.90	100.20	5	100.75	100.10
y+2	5	111.25	111.55	5	111.45	↔
	5	111.25	111.55	5	111.25	71
	5	111.15	111.65	5	111.75	53.45
y+3	5	118.40	119.05	5	57.10	↑

Fig. 2.11 Bid-Ask spread

off-peak, Germany 0–6 (from 00:00 to 06:00) and Germany 20–24 (from 20:00 to 24:00).

Also, lower on the second-to-last column, we have API 2, which is a hard coal quotation at Amsterdam, Rotterdam, Antwerp, and harbours. Next to it, is a high-calorific natural gas quotation TTF (title transfer facility). On the bottom right, we have emission allowances and on the bottom left, we can see recent trades.

The screen shows us the different bids and asks added by the market participants who want to either buy or sell. Similar screens will be used anywhere where energy as a commodity is continuously traded.

2.4.2 *The Bid-Ask Spread*

In Fig. 2.11 (which is a section of Fig. 2.10), we take a closer look at the display to explain the *bid-ask spread*. What can be seen here is also referred to as an order book. This contains all bids and asks for a certain product. Let us first consider the bids submitted by a potential buyer of a commodity, consisting of the bid price and the bid quantity. In the figure, there are three bids and asks for the product "Germany BSLD" (BSLD stands for baseload) for the next year ($y + 1$, highlighted in the orange square) and three for the year after next ($y + 2$). For the first bid in the orange square, the quantity is 5 MW. This means that the potential buyer is willing to buy this product for the year $y + 1$ contract at a maximum price of 100 €/MWh.[10] The bid price always represents the maximum price the buyer is willing to pay for the commodity. In this example, there is one potential buyer who is willing to pay 100 €/MWh, and in the following rows there are two more buyers, who are both willing to pay 99.90 €/MWh for this product.

If someone (for instance at the trading department of a generator) wants to sell, for example, 10 MW of "Baseload Germany", they can first click on the bid of

[10] Note that the total monetary value of this deal would be 5 MW × 8,760 h × 100€/MWh = €4.38 Mio. - potentially executed at the click of a button.

100 €/MWh, and then a new window will ask for confirmation (this confirmation can often be deactivated by the user). Once the transaction has been confirmed, the sale of 5 MW at 100 €/MWh is immediately executed. Then, the generator wants to sell the other 5 MW, but that must be done at 99.90 €/MWh because these bids represent the current interest in the market, and voluntarily one bidder is willing to pay 100 €/MWh, while the other is only offering 99.90 €/MWh.

The bid quantity represents how many megawatts a buyer is willing to buy for the respective price. The quantity entered is typically fixed, with 5 MW being the standard order size. If a market participant wants to sell or buy less than the amount shown on the screen, this cannot be done just by clicking. In this case, it is necessary to ask the broker (e.g. on the phone or by text message) if the bidder, who is currently unknown and has a bid at 100 €/MWh, would be willing to buy a different amount. This often works, but sometimes the other party may refuse, maybe because they prefer to deal with standard product sizes, since it may be easier to resell them later, if necessary.

In summary, if someone wants to sell and is looking at potential buyers on the screen, they will look at the bids, through which potential buyers have expressed their interest. Then, the seller only needs to click on the screen to complete the transaction. Alternatively, the seller can also submit an ask (shown in Fig. 2.11 in the blue square). The asks offer from potential sellers and by clicking on them, a market participant can initiate a purchase. If someone clicks on a bid, they sell. If they click on an ask, they buy.

Like a bid, an ask consists of a price and a quantity. The ask price represents the minimum price that the seller is willing to accept for the commodity, and the ask quantity represents the exact quantity that the seller is willing to sell at that price. In this example, someone is willing to sell at 100.10 €/MWh and if potential buyers are willing to pay that, they only need to click on the ask and the purchase is completed. If a buyer wants to buy more, they will need to go to the next available ask at 100.20 €/MWh and pay a little more.

The information on bid and ask shown in Fig. 2.11 is also a measure of market liquidity. First, it indicates how much energy can be bought at moderate or low price changes. The higher this amount is, the higher is the market liquidity. If the buy or sell side of a market contains only one bid and nobody else is willing to sell, then this market is illiquid. In the example above, the market is more liquid, because a total of 15 MW can be sold with a drop in prices of only 0.10 €/MWh (from 100.00 to 99.90 €/MWh). Furthermore, there may be additional bids as the screen shows only the best three in this configuration. Note that these entries, i.e. all quotes on a specific product available on any given point in time, form the so-called *order book*.

Second, the price difference between the highest bid and the lowest ask is also only 10 cents (100.00 to 100.10 €/MWh). This means that if for some reason someone would buy and immediately sell again, they would only lose 0.10 €/MWh.[11] These

[11] This could be done by clicking on the 100.10 €/MWh ask on the right and then on the 100.00 €/MWh bid on the left. Of course, this is only a way to measure liquidity and does not make sense in practice.

ten cents are called the bid-ask spread. The more liquid the market is, the less transaction costs participants generally have. For a more in-depth analysis of liquidity, the reader may also refer to Sect. 2.6.

To sum up, the bid-ask spread is defined as the amount by which the lowest ask price exceeds the highest bid price for a commodity. It is a central indicator of the liquidity of an asset: The smaller the spread, the higher the liquidity. A trade or transaction occurs when a buyer in the market is willing to pay the best offer available or if the seller is willing to sell at the highest bid.

There are hundreds or even thousands of traders sitting in front of their screens watching the market. They can either enter and submit their bids and asks on the platform and these will immediately appear on the screen, or they can initiate a deal by simply clicking on what somebody else submitted. Note however that traders need to be registered to do this.

2.4.3 Pay-Off Profile of Long-Term Contracts

In energy markets, the price of a long-term contract is determined on platforms as discussed above. But what is the driving force, guiding market participants' bids and offers? For any long-term contract, bids and offers of all market participants (both buyers and sellers) are driven by one factor: The expected spot price in the delivery period. All long-term contracts are a bet on the expected spot price (plus or minus a risk premium[12]), because every market participant could alternatively wait and sell or buy the electricity on the spot market. Hence, the price of a long-term contract reflects the market participants' expectations of the spot price during the delivery period. If the market expects high prices in the future spot market, the price of the long-term contract will be high, and vice versa.

Note however that we refer to expected future spot prices during the delivery period of the contract. In contrast, today's spot price is hardly relevant to the price of long-term contracts currently traded. These are completely different products.[13]

Lastly, we present the pay-off profile of an exemplary long-term contract. Assume someone has bought a long-term contract for 25 €/MWh, and the contract is now approaching delivery and traded on the spot market. In Fig. 2.12, the horizontal axis shows the spot market price in a range from 0 to 50 €/MWh, and the vertical axis shows the resulting profit or loss of the buyer. If the spot price is 25 €/MWh, the profit

[12] The literature agrees a risk premium should be included, but it does not agree on the sign.

[13] Nonetheless, there is some weak correlation because today's electricity system does influence the future's electricity system, e.g. due to long investment cycles. For instance, if we excess supply today, supply may tend to still be excessive e.g. next month or next year. However, in terms of renewables, we need to distinguish further between short- and long-term price effects at this point. Take the example of expecting an abundance of renewable energy tomorrow. This will result in low or even negative prices on the day-ahead market. However, this is meaningless for the front-year price because the weather conditions for next year are not yet known. Hence, we do not know how much renewable generation we will have, so we must work with forecasts and expected averages.

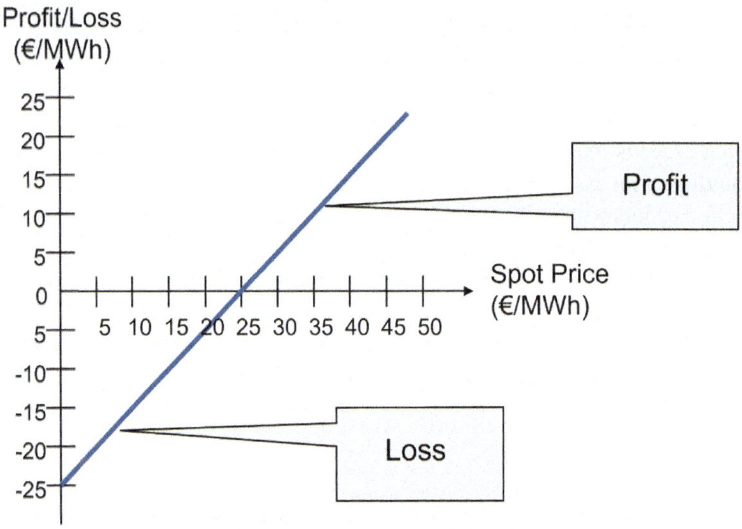

Fig. 2.12 Payments profile of a forward/future purchase (Long)

is zero, because the product has been bought for 25 €/MWh and alternatively could now be bought on the spot market for 25 €/MWh (or resold at this price without a loss or profit). In this case it does not matter whether the forward was bought or not, because the price is the same.

If the spot price turns out to be 50 €/MWh, then it is reasonable to say that it was a good deal: Something that is now worth 50 €/MWh has been bought for 25 €/MWh. This could for instance be a front-year contract. When the front-year goes into delivery and the spot market for that product opens, the buyer can choose whether to consume the product himself or to sell the electricity on the spot market for 50 €/MWh.

On the other hand, if the spot price is very low and turns out to be zero, then the buyer has paid 25 €/MWh for something that is now worth zero, and he thus has a loss of 25 €/MWh. We can do the same analysis for all possible spot prices: If the spot price is above 25 €/MWh, the buyer makes a profit, if it is below, a loss. This is called the payout profile of a forward or a future.[14]

[14] With forwards and futures, the buyer has the right and also the obligation to take the commodity. In contrast, with options the buyer can choose whether or not to exercise them. Options are explained in more detail in Sect. 3.3.3.

2.5 Forwards, Futures and Non-standardised Products

Long-term contracts can be traded in the form of different products. The two most common ones are forwards and futures. Hence, this section will first introduce forwards and futures and explain their commonalities and differences. For reasons of completeness, we will also cover non-standardised products. As forwards and futures are typically traded on different trading places, we will finish the section by discussing energy exchanges, broker platforms, and over-the-counter markets.

2.5.1 Forwards and Futures

Long-term contracts, used to buy and sell energy commodities on continuous trading platforms, can be divided into *forwards* and *futures* contracts. While the two types of contracts have noteworthy differences, we would like to first point out that they have near-identical payout profiles, and in many practical implications, buying a forward is like buying a future.

Turning to the differences, futures are usually traded on exchanges and are standardised exchange products. Forwards on the other hand are mostly traded on broker platforms.

Another difference between futures and forwards concerns price settlement. In the case of forwards, payments are usually made during or shortly after the delivery period. This is because forwards are settled according to negotiated rules and these rules are often based on framework agreements like the European Federation of Energy Traders' (EFET) framework agreement. This agreement specifies that payments for the previous month are settled on the 20th of the following month. This means that if a front-year contract is signed today, no payments are made initially. If EFET rules are specified in the contract, money will be transferred for the first time on the 20th of the month following the month of first delivery. Signing a front-year contract, this would be the 20th of February of next year. Most individually negotiated forwards have either this or similar agreements.

In contrast to that, in the case of futures, financial flows with the power exchange start immediately, because the power exchange requires so-called margin cash flows. These payments cover the difference between the agreed contract price and the current market price of the contract. This means that after a future has been traded, daily price changes are compensated between buyer and seller. In this way, the exchange, which is counterparty to both buyer and seller, avoids losses if a market participant becomes insolvent (i.e. goes bankrupt).

Assume for example a company buying a future for 50 €/MWh and the price falls to 30 €/MWh. Without the daily settlement via the company's margin account, the exchange would be taking a risk: If this company cannot pay what it promised (50 €/MWh) due to bankruptcy, the exchange takes over the commodity. However, the value of the contract has declined to only 30 €/MWh and the exchange can therefore only sell it at this price. The exchange incurs a loss. To avoid the possibility of such losses, the difference between 50 and 30 €/MWh is the payment that the exchange deducts from the market participant's account on at least a daily basis. If that company goes bankrupt, the exchange can cover its loss of 20 €/MWh from the company's margin account. Consequently, the exchange is only exposed to limited financial risks and is therefore unlikely to go bankrupt itself.

Another difference between forwards and futures is that forward contracts are usually settled physically, while futures are settled financially. Therefore, the future is compared to and financially settled against the spot price. Forwards, on the other hand, typically include physical delivery of the commodity to a specific location. If you buy an energy forward and keep it, then when the delivery date comes, the energy commodity will be physically delivered. Many actors are interested in physical delivery. Municipal utilities, for instance, are ultimately obliged to supply their customers with electricity and natural gas. Hence, they need physical delivery if they do not want their customers to be sitting in the dark or in the cold. Additionally, a company that generates electricity or natural gas needs a physical instrument, because it needs a buyer to handle the delivery of the physical commodity. Other companies, e.g. trading houses, investment banks, and hedge funds, do not require physical delivery. Note that this does not imply that the latter only trade futures while the former trade forwards. Typically, other aspects dominate this decision as it is usually easy to convert a financial contract into a physical contract and vice versa. Nonetheless, contractual obligations need to be managed carefully.

These differences also result in small price differences between forwards and futures. For example, the differences in cash flows cause variations in interest rate payments which must be considered: the buyer of a forward can keep the money for a longer time. However, the analysis in this book focuses on the similar cash flows: If the buying price is 50 €/MWh, and the spot price is 70 €/MWh, the payout is 20 €/MWh regardless of whether a future or a forward was traded. The profit or loss of both contracts thus results from the difference between the forward or future price, F_h, and the spot price, S_h for a delivery hour h, i.e. $S_h - F_h$ for a long position and $F_h - S_h$ for a short position.[15] In the end, the comparison with the spot price determines whether a deal has been worthwhile or not. Note that for this reason we will use the term forward for long-term contracts with this payment profile, i.e. synonymously for both forward contracts and future contracts, in the remainder of this book.

2.5.2 *Non-standardised Products*

Non-standardised products are tailor-made to fit the needs of a single market participant. An example could be a complex option contract financially replicating the

[15] The terms *long* and *short* are explained in detail in Sect. 3.1.1. In brief, a long position means that the forward was bought and a short position means that it was sold.

pay-off structure of a specific power plant, taking into account the fuel type, the efficiency and potentially even individual outage times. Another example is a specific demand schedule for an industrial consumer, specifying hourly demand, potentially also including site downtime. Such contracts cannot be traded on trading platforms because they lack liquidity—in the extreme, a specific contract may only be traded exactly once.

An important class of non-standardised products is thus a *schedule*, which we define as an energy time series. This can be for instance the 96 periods of 15 min of the following day or the 8760 hours of a year.[16] Typically, the values in a schedule vary, e.g. following a demand profile.[17] Trading a schedule thus requires an exchange of the time series data in question, e.g. in CSV format.

In such cases, it is optimal to engage in bilateral OTC trading. In the abovementioned example of a power plant operator, they could contact an investment bank or a tier 1 energy company and negotiate an individual deal beneficial for both sides. We consider the trading of hourly electricity schedules as non-standardised product in this book, even though some broker platforms enable the upload of schedules. However, they mostly have comparably low trading volumes. Note again that non-standard products may become standard products and vice versa, reflecting and following the needs and interests of market participants.

2.5.3 Trading Places

There are different places to trade energy commodities that specialise in trading different products. Figure 2.13 shows the properties of the three main trading places and their characteristics.

Exchanges are regulated entities, which are standardised and supervised by the state. For instance, the European Energy Exchange has a supervising authority, which is the Ministry for Economic Affairs, Labour and Transport of the state of Saxony, Germany. There is a representative of this ministry in the exchange council taking part in the meetings and supervising all major processes.

Exchanges are a central counterparty for transactions, they trade standardised products, and they provide clearing services (data transfer, confirmation and others). The trading process is performed electronically and only members can trade on the exchange. Non-members who want to carry out a transaction must pay a commission to a member to do so. Additionally, there are fixed transaction costs, i.e. buying and selling megawatt hours on these exchanges have a fee. The fees serve to cover the exchange's costs. All transactions are fully anonymous because the exchange

[16] Throughout this book—unless otherwise stated—, we assume a year consists of 8760 hours, i.e. is a regular year with 365 days. In reality, of course, a leap year comprises 8784 hours and a base product for a leap year covers 366 days and thus specifies a delivery of 8784 hours.

[17] Note however that standard product delivery can also be interpreted as a schedule. For example, a 5 MW base contract for the front year specifies a schedule for the 8760 hours of that year, i.e. an energy delivery of 5 megawatts in each hour of the year.

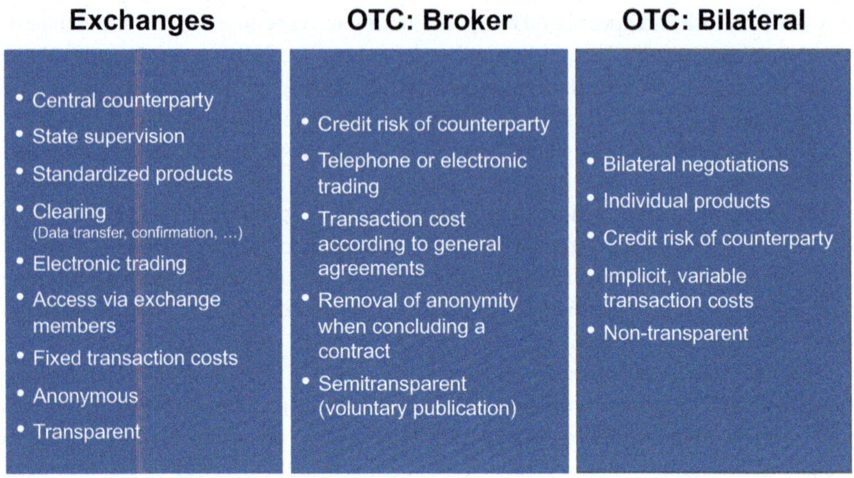

Fig. 2.13 Trading places

serves as the central counterparty. Due to legal requirements the exchange must be transparent; i.e. it is obliged to provide all sorts of information.

A lot of information on exchanges can be found on the websites of the respective energy exchange. Some of this information is free, particularly information about recent market activity. Other information, such as disaggregated historical time series of traded volumes and prices, is often paywalled.[18] Take the website of the European Energy Exchange (EEX) as an example.[19] You can access the market data section and, for instance, retrieve the most recent quotes for the front year of base electricity contracts for Germany or any other European country of your interest. Also, you can look for the price of the front-year base contract for natural gas, and CO_2 emission allowances. With this information, you can estimate the profit of a gas-fired power plant if it is operated 8760 hours a year and has to pay for CO_2 emission allowances. All that you need is this price information and the power plant's efficiency.[20]

The exchanges provide one way to buy a certain electricity product, e.g. the front-year base product. However, a product with the same pay-off profile can be purchased from an OTC *broker platform*.[21] Typically, such a platform is organised as a web application. This is how most volumes are traded with brokers. In addition, the broker can also be contacted by speaker phone (typically causing some background noise on the trading floor) or via instant messaging. Trading on the phone can be particularly

[18] The exchanges have found out that this information is valuable, so they charge for the right to use it.

[19] http://www.eex.com.

[20] Luckily, gas-fired power stations do not run 8760 hours per year, but only during the most profitable hours, when power prices are above the base price. As a result, they can still be profitable even if their variable cost is above the base price.

[21] Historically, term "over-the-counter" meant that two persons meet and agree on terms of a business deal over-the-counter, for example in a bank.

helpful for less liquid products. The broker can be informed about the desired product on the phone, inquire about it and find a way to get a quote. This often helps when a trader requires a non-standard order size (e.g. just one MW instead of the five MW standard offer).

Four aspects should be considered for the decision of whether one wants to trade on broker platforms. First, trading on broker platforms leads to credit risk with counterparties: (in contrast to an energy exchange) there is no central counterparty. Instead, both counterparties' identities are disclosed after the transaction at the latest. For instance, if a municipal utility wants to buy electricity from a power plant operator, the broker's function is mainly to bring the two parties together (i.e. to broker in the pure sense of the word). The broker provides the market platform and makes the deal feasible. He is, however, not involved in the deal itself as the forward contract is signed between the utility (buyer) and the power plant operator (seller). If one of the counterparties goes bankrupt after the deal has been signed, the counterparty may face losses. This is referred to as credit risk and will be further elaborated in Sect. 4.3.

Second, transaction costs are negotiated between broker and client, which means they may be more flexible than fees on exchanges. Furthermore, they often differ from broker to broker.

Third, transactions on broker platforms are not fully anonymous. While market participants do not initially know which company submitted the bids and offers on the broker platform, counterparties are disclosed as soon as a quote is accepted. This is different on an exchange, where all deals have the exchange as central counterparty. Anonymity can therefore be preserved.

Fourth, OTC broker platforms can be less transparent than exchanges because they are less regulated. From the perspective of market participants, this has advantages and disadvantages: There is less bureaucracy, but at the price of reduced transparency.

The final trading place worth mentioning in our context is *bilaterally negotiated* transactions. These types of deals can be used, for instance, if a utility wants to buy or sell individual contracts—essentially anything with inherently low liquidity. Mostly, this refers to the non-standardised products discussed in Sect. 2.5.2.

To discuss another example, consider a baseload electricity delivery for the next 20 years. There is no liquidly traded product for this contract on the exchange, and broker platforms are unlikely to provide quotes either. However, it is possible to negotiate an OTC transaction bilaterally. Mostly, such negotiations are not performed by trading departments but by an origination department or another specialised department of the company, which also assesses the risks and opportunities associated with the transaction. Both counterparties must negotiate and agree upon all the details of this individual contract, including, e.g. how, when and where the electricity will be delivered, what legal framework is selected and of course at what price the deal is settled.

In a bilateral negotiation, there is of course always a credit risk for the counterparty. Moreover, in bilateral negotiations, transaction costs are fully negotiable. Most of the time, they are not even mentioned explicitly but implicitly covered by the conditions of the contract. Despite this, transaction costs may be high due to the complexity of the negotiation process. For example, if the negotiation for a given contract takes one week and involves several people (assessing the terms of the contract, calculating the appropriate price for a potentially complex product, phrasing a water-tight contract) then these costs are implicit in the deal.

Deals resulting from bilateral negotiations are completely non-transparent: Whenever two counterparties sign a bilateral contract, the public does not get to know the details or even that the contract has been signed at all. It remains company secret and does not have to be reported.[22]

2.6 Liquidity

Sarr and Lybek[23] define five characteristics of liquid markets: "(i) tightness; (ii) immediacy; (iii) depth; (iv) breadth; and (v) resiliency. *Tightness* refers to low transaction costs, such as the difference between buy and sell prices, like the bid-ask spreads in quote-driven markets, as well as implicit costs. *Immediacy* represents the speed with which orders can be executed and, in this context also, settled, and thus reflects, among other things, the efficiency of the trading, clearing, and settlement systems. *Depth* refers to the existence of abundant orders, either actual or easily uncovered of potential buyers and sellers, both above and below the price at which a security now trades. *Breadth* means that orders are both numerous and large in volume with minimal impact on prices. [...] *Resiliency* is a characteristic of markets in which new orders flow quickly to correct order imbalances, which tend to move prices away from what is warranted by fundamentals. These terms reflect different dimensions of the extent to which an asset quickly and without significant costs can be transformed into legal tender". Hence, *liquidity* answers the question of how quickly and at what cost one can open and close positions in a specific product.

Note that we have already discussed in Sect. 2.4.2 the bid-ask spread as a crucial measure of liquidity: the lower the bid-ask spread, the more liquid the market. Furthermore, we have discussed the order book, i.e. how many bids and offers for a specific product are present in the market at any point in time—and how close together their quotes are. This measures the depth and breadth in the definition above.

Of course, the paradigm is that more liquidity is better. The ease of opening and closing positions in a timely manner while avoiding larger losses makes trading more mature and beneficial. Furthermore, it becomes easier and cheaper to deal with risk in this case. For example, if you are investing in a wind farm, you already know that you will have a long position even far in the future, because you assume that your wind farm will generate electricity for at least up to 20 years after it is built. From a risk management position, it would be beneficial to be able to sell some of this energy

[22] Note however, that in most cases, this is how it should be as the terms of the contract are business secrets. The harm suffered by the two parties when details are disclosed often outweighs the public interest in information disclosure.

[23] Lybek and Sarr (2002, p. 5).

(more about this in Chap. 4). The more liquidly traded the associated products are, the easier it is to properly manage the risks.

However, to discuss the concept further we continue the example of a wind farm and discuss a lack of liquidity first: unfortunately, the product (volatile generation from a wind farm far in the future) is not very liquidly traded and thus it would likely be sold with a discount. Nonetheless, assume the investor wants to sell a forward contract with delivery during calendar year 2040. At this point, the best option to sell this contract is as a non-standardised contract on the OTC market. The investor might now approximate the expected annual generation of the wind farm as a base load contract. Selling this more standardised contract would at least hedge some of the risks associated with the investment. Nonetheless, even a standardised base load contract is not yet liquidly traded for the year 2040. Hence, the OTC market would still be the best option at this point in time.

In 2030 trading of this contract will become possible at the EEX.[24] However, even then the Cal40 contract will be illiquid, i.e. it will be difficult and costly to buy or sell such a contract because few buyers in 2030 will be interested in already buying electricity delivered ten years in the future. The closer to deliver the product is, the more liquidly traded it will usually be since interest on both buy and sell sides increases.

These arguments can be generalised to broader lesson: *product standardisation* is needed to increase liquidity. The key reason is that participation in energy trading is voluntary. Nobody can (or should) be forced to trade. At the same time, the more suppliers and buyers are interested in a certain product, the more liquid the market for this product is and the better it is for market participants in general. Hence, the market needs buyers and sellers interested in the products. As a result, there is a trade-off between the "product fit" (level of customization) and liquidity. For example, the wind farm mentioned above sells base electricity even though it knows that the wind farm will have a more volatile production profile. The base contract is more liquidly traded but has significantly less product fit. The problem with the volatile wind profile is that no one else is interested in that product,[25] which is probably unique for their customer. It would be nice for the wind farm operator to be able to find this product on the market, but the trade-off is that it would be very illiquid. On the other hand, liquidity is much higher for the annual base or peak contract, because both are highly standardised products.

For this reason, both baseload (delivery from 00:00 to 24:00 on all days of the contract) and peak load (delivery from 8:00 to 20:00 on weekdays) are typically used to approximate both a production profile of a power plant as well as a load profile for a customer. The exchanges developed such standard products over time because they try to offer products of great interest to the market participants in terms of liquidity and product fit. After all, the more attractive the product is, the more trading activity

---

[24] Trading for yearly baseload contracts currently starts 10 years before delivery at EEX.

[25] Note that in markets with high shares of renewable energy, new financial products (e.g. power purchase agreements—PPAs) emerge to deal with the specific production pattern of wind and solar power. The market evolves.

the exchange has on its platform and the more revenue it generates. Base load and peak load are thus the established market compromise between liquidity and product fit.

It is clear at this point that different products are traded with different levels of liquidity. Even within standardised products (base load and peak load), liquidity varies. There are some general assessments that can be made in terms of liquidity of different standard products in electricity markets. First, base load contracts are generally more liquidly traded than peak load contracts. Second, Fig. 2.14 shows a list of products typically available in the market, grouped by the time before delivery. On the right-hand side of the figure, we see the liquidity of these products and how it increases as the delivery time approaches.

The most liquid short-term products are the day-ahead products, i.e. the hourly products. These are followed by daily products (e.g. the base product for the day after tomorrow), then weekend products (e.g. the base product for the next weekend) and finally weekly products. In general, the further into the future you look, the less liquid a short-term product becomes. This also applies to medium-term products, where the front month is the most liquid product and the quarters—especially those far in the future—are the least liquidly traded. At the long end of the curve (this applies to all long-term products), the front year is the most liquidly traded product, and annual products further into the future are less liquid. Summer products, which cover the second and third quarters, and winter products, which cover the fourth and first quarters, are usually less liquidly traded than the front year.

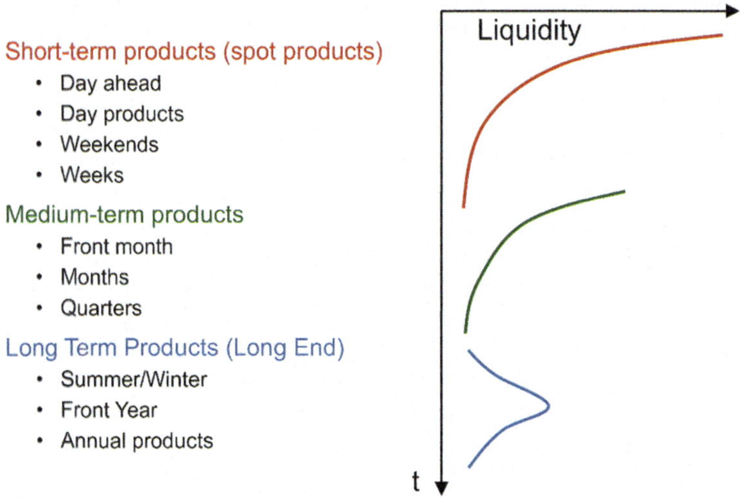

Fig. 2.14 Available products in the market and their liquidity

2.7 Motivations for Trading

Depending on the objective, the motivation to engage in energy trading can be divided into three main groups:

- ***Arbitrage*** achieves risk-free profits from price differences of the same product at the same time on different marketplaces. For example, if you see a very low price on broker platform A, while the price for the same product is higher at broker platform B, you could make arbitrage by buying on platform A and selling instantly on platform B. However, such situations are rare, because many smart and well-paid experts (as well as algorithms and bots) are watching the market and looking for such opportunities. Nonetheless, it can be worth looking for arbitrage, especially in less developed and less liquid markets.[26] There can also be regional arbitrage, for instance, energy is bought in market A and then sold in market B for a higher price. If the price in A plus the cost of transportation rights from A to B is still lower than the price in B, then a regional arbitrage is possible. There are many forms of arbitrage,[27] but they all have the same basic definition: Making risk-free profit.
- ***Speculation*** happens when a position is opened in anticipation of a certain market price development. For example, if you observe that the electricity price for the front-year contract has recently increased significantly, you may believe that the realisation of the spot price will be lower and the current price of the contract is thus too high. On this ground, you may decide to sell a front-year future now (i.e. at the price you think is too high). If your assessment was right, the price decreases approaching delivery and the trade has been profitable. Otherwise, you make loss on the trade. This is the reason why this trade is called speculative: It is based on the belief that someone is smarter than the market. Forward market prices are the consensus of all market participants on the expected spot price (plus or minus a risk premium) and are therefore considered a good estimator. One must be quite confident to expect profits from this type of trading (or actually have information that the other market participants do not have, which is extremely rare). In a nutshell, Fig. 2.15 shows what speculative trading is all about: buy low, sell high!
- ***Hedging*** means reducing the risk of an open position by entering a counter position. Consequently, the two positions are balancing each other. In the best case, for example when a base contract was sold to an industrial consumer, a base contract can be bought, and the respective risks cancel each other out exactly. In other cases, such a perfect hedge is not possible, but the hedge can still be useful if it

[26] Note that this is a reason that some specialised traders prefer less liquidity. While this is perfectly reasonable from a business perspective, higher liquidity increases social welfare from an economic perspective—at least ceteris paribus.

[27] Another example in electricity markets is product arbitrage: for example, if the weighted average price of the four quarters of a year is lower than the price of the whole year, a profit can be made by buying the quarters and selling the whole year. This works with all types of combinable products, e.g. months and quarters, days and weeks.

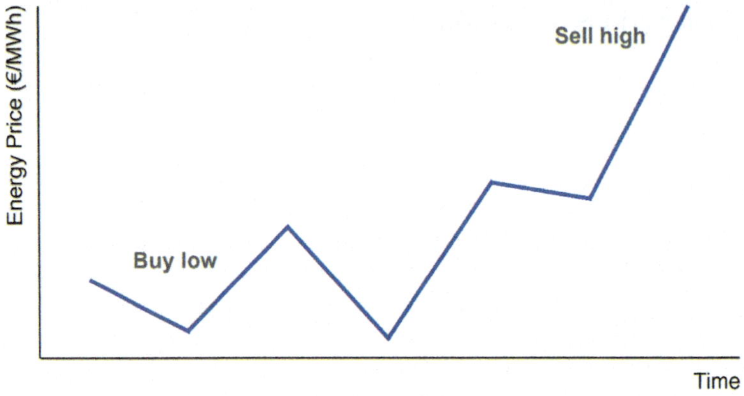

Fig. 2.15 Speculative energy trading

is at least highly correlated to the open position. The risk will at least be mostly reduced. As an example, a power plant may be hedging its natural long position in electricity with a short position in the wholesale market (and similarly vice versa the natural short position in fuel and CO_2 allowances). Further information on the open position and its calculation is provided in the following Chap. 3.

Reference

Lybek, T., & Sarr, A. (2002). *Measuring liquidity in financial markets* (Working Paper No. 2002/232).

Chapter 3
Portfolio Management

This chapter will discuss portfolio management in energy markets. The discussion will present various angles, reflecting the perspectives of companies engaged in energy trading. For that purpose, this chapter is divided into five sections.

In the first section, we will discuss proprietary trading. While it is an art to do it profitably, it is the simplest form of portfolio management in our analysis as it is purely focused on opening and closing positions on the wholesale market. In the second section, we will analyse an energy utility with a retail focus, i.e. energy is bought on the wholesale market but mostly sold to retail customers who consume the energy. The section discusses the different types of customers and contracts, and procurement strategies. In the third section, we will move to an energy utility with a generation focus, and learn concepts such as spreads, swaps, optionality, hedging and marketing flexibility. In the fourth section, we will combine wholesale markets, generation assets and the retail segment to study truly integrated portfolio management. This describes the situation faced by tier-one energy companies, large municipalities and large, energy-intensive industrial companies. Here we review the open position of an integrated company and show advanced financial products that can be used to manage integrated portfolios. We will also give a practical example of what a day on the trading floor looks like. In the fifth and last section, we will demonstrate the application of performance measurement to identify the sources of a company's success or shortcomings, both for the company as a whole and within a company.

3.1 Proprietary Trading on Wholesale Markets

Proprietary or just "prop" trading provides a good starting point for portfolio management in energy markets because—compared to the alternatives analysed later—less elements are involved. For example, prop traders neither require physical delivery nor

F. Müsgens and A. Bade, *Energy Trading and Risk Management*,
https://doi.org/10.1007/978-3-031-57238-8_3

do they operate physical assets such as power plants. The prop trading approach is not purely theoretical, however, because there are many companies, such as investment banks, who do not own generation assets but still trade actively on energy markets.

As we discussed before, proprietary trading is usually done by trading houses, hedge funds and investment banks. Proprietary trading strictly speaking means that the trading is not done on behalf of a client and not to supply a retail client. While investment banks often engage in trading activities on the behalf of clients, we stick to the narrow definition where profits and losses are fully taken by the prop trader. There is a beautifully short definition of proprietary trading from the Nasdaq website: "Principal trading in which a firm seeks direct gain rather than commission dollars".[1]

3.1.1 Open Position of a Prop Trader

A key concept in this context is the trader's *open position*. The open position drives chances and risks and is therefore crucial in all prop trading activities. The open position of a prop trader is the difference between what has been bought and what has been sold for a certain delivery period *h*. Note that while the delivery period can be any time interval, it is useful to analyse the open position either in hourly or in quarter-hourly resolution. Using an hourly resolution, the formula in Eq. (3.1) calculates the open position (OP) for all 8760 hours of the front year:

$$OP_h = BUY_h - SELL_{h.} \tag{3.1}$$

Before any contract is signed, the prop trader's open position is zero for each of these 8760 hours. Now let's assume that the trader buys a 5 MW front-year base contract. This changes the open position for the first hour of the year and for all subsequent hours of that year. For all hours of the front year, the trader has bought 5 MW and has sold nothing, so the open position is + 5 MW for all hours.

When the open position is positive it is usually called a *long position*. Being long thus means that one has bought more than one has sold. The open position can become negative. This simply implies one has sold more than one has bought. This is called a *short position*. Hence, the open position can be positive, negative or zero.

The open position for a delivery period *h* will change over time during the trading period, (again, it is important to know the difference between these two periods, see Sect. 2.1.). For a prop trader, this can happen through transactions, as well as due to changing expectations for complex financial products, like options. If you sold an option, then the open position can change because, depending on the price, the expected sale varies.[2] The process of managing the open position is often referred to as portfolio management. This is the core of revenue generation and thereby at the heart of the analysis in this chapter.

[1] NASDAQ (2023).

[2] More details about the way options work will be shown in Sect. 3.3.3.

3.1.2 Balancing Groups—Trading and Physical System Stability

As the prop trader pursues purely financial interests, the trader is not concerned with the physical stability of the system. He is not even interested in the physical delivery of any energy. Hence, a prop trader's open position is necessarily zero at the start of any contract's delivery period. At the same time, the trader is active in a complex and sensitive system—the energy system. Every actor in this system, including the prop trader, is required to contribute to system stability. For the prop trader (and all other traders), this is done stating their open position for the next day on the day before delivery. The underlying process is organised with the concept of *balancing groups*.

The concept of balancing groups is an essential tool linking the financial and the physical world of electricity. We have already pointed out that electricity is a hardly storable commodity. Hence, generation and consumption must be equal in real time to guarantee a stable grid frequency of 50 Hz.[3] Large deviations from this target frequency can damage some electrical equipment or even bring the entire system to a standstill. Maintaining a stable frequency of 50 Hz is the responsibility of the transmission system operators (TSOs).

If all market participants were just trading energy contracts, including those with physical delivery, without reporting anything, the TSOs would not know who produces and consumes electricity in their grid. Avoiding this is essential because otherwise the TSOs could not guarantee a stable frequency and a safe operation of the system.

Consequently, they gather the required information from balancing groups. Each producer, consumer and trader of energy contracts needs to be part of such a balancing group. On the day before delivery after the day-ahead auction,[4] each balancing group must be balanced (in expectation). An investment bank that trades electricity contracts must thus by law be part of a balancing group. The balancing group must prove to the respective TSO that the balancing group indeed is balanced, which implies that the investment bank bought and sold the same amount of electricity in each of the 24 hours[5] of the next day. This requirement enables TSOs to provide a stable physical operation of the system: firstly, at least in expectation the whole system is balanced as the system is essentially the aggregate of all balancing groups, which are all individually balanced. Secondly, the balancing groups' nomination contains information on contracts, consumption and production.

Figure 3.1 shows this relationship between the physical world (top of the figure) and the financial or "schedule" world (bottom of the figure). The physical energy world is highly complex and consists of millions of kilometres of grids, thousands

---

[3] This holds in the European grid and many other grids in the world. In the USA, for instance, the target grid frequency corresponds to 60 Hz.

[4] In Germany and in most other European countries, this needs to be done until 14:30.

[5] To be more precise, for all 96 periods of 15 min of the next day as balancing schedules in Europe are organised in quarter-hourly resolution.

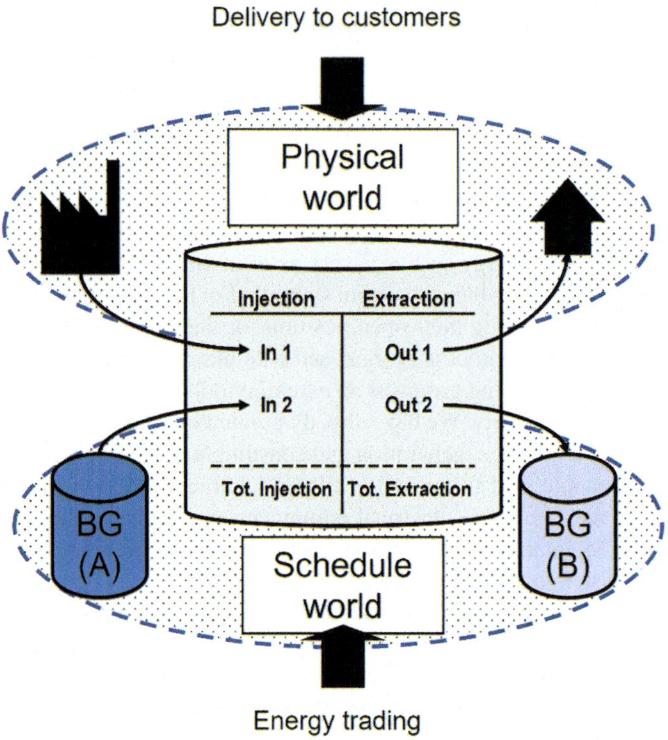

Fig. 3.1 Relation between the physical world and the financial world

of generators and millions of consumers. Electricity is essential to modern life and has a complex physical environment.

However, these complex and sensitive physical systems have been combined with the benefits of liberalised energy systems, significantly increasing efficiency, partly due to professional energy trading and risk management. Liquid energy trading has not brought the systems closer to collapse. While there have been problems in some liberalised systems, these are typically not related to the aspects discussed here. All in all, the interaction of the market and the physical reality works surprisingly well. From an economic perspective, this has brought a lot of advantages in terms of efficiency.

In the financial world, balancing groups ensure that consumption and feed-in into the grid are balanced on the basis of financial contracts. A balancing group can comprise several, one, or no grid users. A balancing group without physical grid users (such as power plants or consumers) is called a "trade balancing group" and is solely used for trading purposes. Such balancing groups are suitable for prop traders who are uninterested in physical delivery.

3.2 Energy Utility with a Retail Focus

In this section, the focus will be on retail customers. The main difference to pure prop traders is that an energy company with a retail focus earns revenues from delivering energy to customers. This energy needs to be procured on the wholesale market. In contrast to pure prop trading, making profits by buying and selling at the right time is often still an objective, but not the single "raison d'être" anymore. In this section, we assume for didactic reasons that the wholesale market is the only source of energy for the utility company, while the retail customers are the only consumers of this energy. This means that everything bought on the wholesale market is sold to the retail customers.

Note that there is a strong analogy between an energy utility delivering energy to consumers and a large industrial company buying its own energy demand on the wholesale market and consuming it in production processes. In particular, the open position and the procurement strategies described in this section are directly transferable and hopefully useful in that setting as well.

From the perspective of a retail utility, several aspects of the business process must be considered since they all have an impact on the profitability of the company. First, the utility needs to identify possible customers and convince them to buy their energy from the utility—at the best price possible from the perspective of the utility. Second, the utility must buy this energy on the wholesale market.[6] Note that this energy can be bought before or after it is sold to the final consumer. In the first case, the company is long between buying the energy and selling it to the consumer, and in the second it is short between selling it to the consumer and buying it on the wholesale market. Either way, the utility must have bought and sold the same amount at the beginning of each delivery period, as was demonstrated in the last section on balancing groups.

The procurement process is not the same for every company and depends on the company's risk concept. Furthermore, we will learn that this process is influenced by the type and size of the customers, as well as by the amount of energy sold. In this section, we will analyse several procurement strategies as well as the impact they have on the open position.

In the following, it is assumed that the energy utility buys energy on the wholesale market and sells this energy on the retail market. Consequently, we need to understand how the retail market works. This requires that we distinguish between different types of retail customers and different types of energy contracts.

3.2.1 Customer Types in the Retail Market

In the retail market, customers are usually end-users of energy. They do not buy the energy and then resell and rebuy it again. We thus define a retail customer as a person,

[6] In this section, we assume that the utility does not own generation assets.

company or entity that physically consumes energy and buys it from an intermediary such as a utility.

3.2.1.1 Household Customers

Most customers in the retail market are *household customers*. They are characterised by low energy volumes and consequently a low absolute revenue per customer. Another typical property of household customers is that there is often an uncertain contract duration. In liberalised markets, they have the right to terminate their contracts on relatively short notice, which can be challenging for portfolio management.

Additionally, household customers are still rarely measured in real-time. Often, they have annual metering of energy consumption. Their hourly demand profile is thus unknown.[7] This is a challenge as we have already pointed out that "electricity today" is a different product than "electricity tomorrow" (and electricity now is a different product than electricity in one hour). Therefore, the true monetary value of a household's demand profile depends on how much of the annual consumption was consumed during the night, for example, and how much during the day.

As this data is regularly missing, there is the concept of "standard load profiles".[8] A standard load profile represents the average (quarter-)hourly power consumption of a typical household. Standard profiles differ from country to country, from region to region and they may look different in summer and winter. However, one typical assumption of most standard profiles is that there is much less consumption during the night than during the day. Essentially, the standard load profile is an artificial average structure which is used to distribute the overall annual consumption to the hours of the year. This standard load profile can be used by utilities to procure the right schedules for their household customers—without needing to worry about their actual real-time consumption. In this case, the utility uses standard load profiles for day-ahead nomination. The balancing group is considered balanced when the hourly energy amounts that were procured match the standard load profile of all its household customers.

Finally, household customers typically pay a flat price per kWh plus a monthly fixed price fee. This means that the customer always pays the same amount for each additional kilowatt-hour consumed, independent of the hourly wholesale market price in the hour it was consumed. This is a consequence of lacking real-time metering. This, of course, does not create an incentive for households to save energy when it is expensive on the wholesale market: they always pay the same price, even if the wholesale price for that particular hour spikes.

Given the nature of the household customer segment (a large number of small consumers), utilities must have efficient and automated processes to make profits

---

[7] This is changing with increasing digitisation, but it will still take time until real-time measurement is the standard for all household customers in Europe.

[8] See e.g. Scholz and Müsgens (2017).

selling electricity to households. For example, it is typically unprofitable to send a sales representative to a customer and negotiate the individual terms of a contract. Nowadays, it is normal to offer standardised contract conditions via the internet. Ideally, the customers will find the utility's website (or the website of a service provider), provide their personal information and maybe sign the contract digitally. During this process, the customer has not once seen a sales representative of the utility in person. Obviously, this is done with the aim of reducing costs, because ideally the main goal for customers is to get the best tariff. At the same time, it also saves time and work on the utility's side.

The fact that the contract duration of households is uncertain is an implicit risk for portfolio management. Typically, the electricity to supply households is bought on the wholesale market before delivery, because the utility needs to be able to give an offer to the customers and specify the contract price, which depends on the wholesale price. At this stage, however, the utility does not know how much energy will have to be purchased, as it is not known how many of the existing household customers will remain customers, nor how many new customers will be added.[9]

This implies a volume risk for the utility. It needs to make assumptions about both the retention rate of existing customers (maybe 80%) and the acquisition rate of new customers. These assumptions may be correct, but if not, some additional trades will be necessary on the wholesale market. For instance, if instead of 80%, 90% of household customers remain, the utility needs to buy additional electricity on the wholesale market. If it is only 70%, it will have to sell the corresponding amount. The same accounts for incorrect assumptions about the acquisition rate.

3.2.1.2 Large Industrial Consumers

In addition to households, there are also large industrial customers in the retail market. In contrast to household customers, they have relatively high energy consumption and generate high absolute revenues per customer. Therefore, utilities have sales representatives who interact with these clients and negotiate individual contracts. Since such contracts can have a volume of several million euros, it is worth spending a few hours negotiating or visiting clients several times to agree upon the optimal terms of a contract. Thus, the retail business is not just about having the best processes and keeping costs down, but also about the marketing and sales team convincing the customer to sign a contract.

Another property of large retail customers is that they have specified delivery schedules, possibly with some flexibility, such as plus or minus 10% deviation per hour. The deviation of an hour from the previously agreed schedule is often valued at the day-ahead price of that hour on the wholesale market. This requires real-time load measurement. The necessary equipment is usually in place for large industrial customers.

---

[9] Müsgens and Steinhausen (2010).

3.2.1.3 Small and Medium-Sized Enterprises

Between the households and the large industrial customers, there is a continuum of other retail customers, mostly so-called small and medium-sized enterprises (SMEs). Depending on their size and sector, they are more like household customers or more like large industrial consumers.

3.2.2 Contract Types in Retail Markets

There are two types of contracts between energy suppliers and retail customers: Fixed volume and open volume contracts.

3.2.2.1 Fixed Volume Contracts

In this type of contract, the delivery structure is predefined and fixed when the contract is signed. The delivery structure can either be an individual load profile (i.e. a schedule) or a standard product (i.e. baseload or peak load).

For example, an aluminium producer can plan its demand schedule for the next year in advance. As defined, such a schedule contains specific hourly electricity demands for the 8760 hours of the next year. In practice, this will result in a spreadsheet table with the 8760 hours for the coming year and the specified amount of electricity to be supplied in each hour. This way, both sides have agreed on the precise delivery schedule before the delivery period starts. For that reason, the utility selling this schedule to the aluminium producer knows the exact amount of energy sold and could also buy the exact same schedule on the wholesale market.[10] This would eliminate any price risk.

3.2.2.2 Open Volume Contracts

Open volume contracts retain a high degree of flexibility for the customer. Most customers do not know exactly how much electricity they are going to need on a specific day and in a specific hour of the following year. Therefore, signing a contract with a predefined fixed hourly delivery schedule leaves them exposed to market risk and they eventually need to buy or sell deviations between realised hourly demand and scheduled demand. Quite often, it is therefore considered beneficial for both sides to sign a contract in which the utility guarantees to supply the amount of electricity required by the customer without a quantity limit or specific bandwidth.

---

[10] Note again that such a schedule is a non-standardised product and is not quoted on trading platforms.

However, such contracts obviously create challenges for the utility. The procurement on future markets must be based on load forecasts, as the utility does not know how much a certain customer is going to consume. The utility only has an expected value, and it can buy the expected amount. However, most likely it will end up with a deviation because the purchased quantity does not necessarily correspond to what is ultimately delivered.

Consequently, the delivery profile contains both a volume risk and price risk. Even if the utility organises the delivery of the demand realisation by trading deviations from the planned schedule on the day-ahead market, the question remains as to who pays for those deviations, or in other words, who bears the risk.

When the risk is borne by the utility, it essentially sells a procurement option to the client. The client can buy electricity from the utility, but they do not have to. Options are valuable and the utility needs to ensure that it is paid appropriately for them, as it bears the risk for the customer.[11] For instance, assume that a utility sells an open volume contract where the amount of electricity consumed by the client simply follows their current demand realisation. All the portfolio management team can do to minimise the risk is to buy the expected demand value. When that is done, assume the electricity price goes up and the client consumes more than expected—which is allowed according to the contract—all the portfolio management team can do is to buy the missing amount of electricity on the wholesale market. Obviously, this will cost money. Of course, prices can also decrease, but it is still a risk. Hence, it is best practice for utilities to charge a risk premium as compensation.

However, in the worst case, this risk is systematic: if the client is smart and the contract allows it, the client could "game the contract against you". This can simply be done by monitoring the market price. Since the client knows he has the option to obtain electricity from the utility at a predefined price through the contract, they can simply exercise this option if the current spot price is higher than the contract price and then resell this energy at that higher price. This is possible if the utility agreed to deliver at a fixed forward price. As you can see, the risk in this case is not even stochastic, it is a systematic risk which will be gamed against you.

Consequently, most municipal utilities prefer another type of contract, where the client can vary the volume but also bears the risk. They offer the client to deliver what they need but the client must pay the spot price for the difference between the agreed schedule and the electricity delivered. For the deviation, the utility just acts on the spot market on behalf of the client. The utility does not have a price risk in this setting. This can be an elegant solution for this situation. The risk is with the consumer, who is most able to estimate the demand schedule in the first place and retains the incentive to follow the forecast. At the same time, the consumer does not have to worry about procurement details (e.g. market access) and the operation of balancing groups as this service is provided by the utility.

[11] Grothe et al., (2006).

3.2.3 Open Position of a Utility with Retail Focus

For a utility with a retail focus, we define the open position as the difference between purchases (BUY_h) and consumption by retail customers (CON_h) for a delivery period h, which again is assumed to be one hour in a future year. This open position is shown in Eq. (3.2).

$$OP_h = BUY_h - CON_h(p_h). \tag{3.2}$$

Consumption of retail customers may depend on the price (as indicated by the index p_h). The reason is that they may decide to postpone, for instance, energy-intense processes as a reaction to high prices.[12]

Note that this open position and the following discussion of procurement strategies can be transferred nearly one-to-one to large industrial consumers who manage their own demand portfolio: their open position is the difference between what they bought on the wholesale market and their own expected consumption, just as described in Eq. (3.2).

As we mentioned before, the open position can change over time during the trading period. It can be positive, negative or zero. However, on the day before delivery, all balancing groups must be balanced to help maintain a stable grid and frequency. This implies that the open position must be zero for each of the 24 hours of the next day.

For a prop trader, electricity bought must hence exactly equal electricity sold for each hour (see Sect. 3.1). For a utility with retail focus, the open position must also be zero at that point in time, but its balancing group includes physical consumption and provision, for instance, from forwards. The latter implies that the municipality has bought exactly as much as it expects to supply in total to its consumers in each hour of the following day. In the case of a large energy consumer with its own portfolio management, this implies that the consumer has bought exactly as much as it expects to consume.

As we have previously mentioned, chances and risks are two sides of the same coin, and there is always a trade-off. There is no chance without risk, like in the old saying "there is no such thing as a free lunch". You do not get anything for free, and certainly no money on competitive wholesale energy markets. Therefore, optimal portfolio management always needs to look at both: chances and risks. For a portfolio management team, price-related chances and risks are mostly tied to the open position and price changes. The larger the open position, and the more frequent and intense the price changes, the greater the chances and risks.

With a positive long open position, a rising price is synonymous with a profit, a falling price is synonymous with a loss and vice versa with a negative (or short) open position. The higher the open position in either direction, the greater the leverage. We can manage the risk by managing the open position, but to do this we need to know how big it is.

---

[12] A typical, albeit extreme, example of this is the energy crisis of the winter of 2022/23, where many public swimming pools reduced the water temperature to use less energy for heating.

We defined the open position for a retail-focused company in Eq. (3.2) as the procured amount minus retail sales or consumption. Sometimes, this calculation is straightforward, but sometimes challenges appear:

- We do not know exactly the retail sales volume for $y+1$ of a household customer (or more precisely a portfolio of household customers) or customers with (partly) open volume contracts.[13]
- Monitoring both buying and selling in real-time can be complicated, i.e. data processing systems of energy utilities may be inadequate.

In such cases, it is difficult to manage the open position because we do not know it in real time, but reasonable estimates can be used.

In practice, if we assume that we will have sold a certain number of gigawatt-hours to be delivered next year or two years from now to households, there is no fixed point in time at which we consider this amount to be allocated and we include this as 100% certain sales in the open position. Instead, we can define a procurement period of maybe two or three years, and on each day, our estimation becomes more certain. Hence, on each day we can include a little bit of the total estimated household selling volumes into our open position. Essentially it boils down to assuming that over a procurement period, each day we buy a little bit in a risk-free setting. And since we said that the average price over the procurement period is the benchmark, if the portfolio management performs well, the procurement prices will be below the average.

Again, the difference between industrial customers is that for these, we have a precise date when we can include the sale in the open position. This means that the portfolio management has a clear task, and they can either buy before that and have a long position out of which they can sell to the customer. Alternatively, they can wait and procure later. For households, we do not have that, so we are forced to use estimates and buy a bit every day (or use another strategy). However, if you have households in your sales portfolio, this will almost inevitably lead to an open position. It is just a question of how to estimate it.

There are several strategies for managing the open position. One of them is the back-to-back procurement, which in theory makes the open position equal to zero by ensuring that purchases and contracts are concluded at the same time. Hence, there is no price risk, no chance and no extra profit.

However, in practice there are reasons for taking chances and the associated risks. Such reasons are in particular the following:

- **Positive contribution margins**. These can be achieved through active management of the open position from the portfolio management team. However, not every company shares this view, there are some risk-averse utilities which prefer to focus more on their sales and clients.
- **Customer requests**. This can be due to flexible purchase quantities, which we have already discussed. If we do not know how much a client will consume, it

[13] As we explained, since we do not know the actual volume, we work with probabilities.

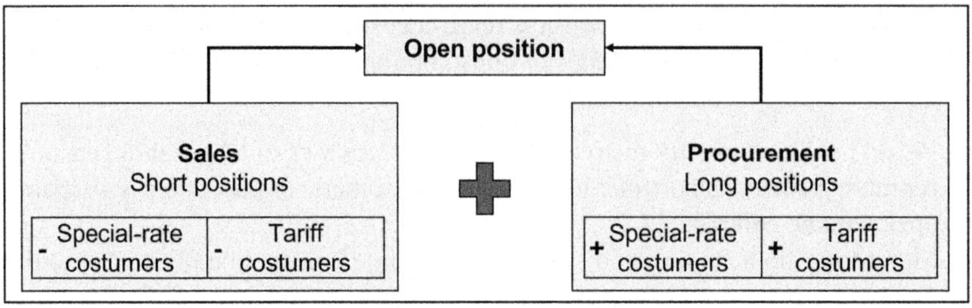

Fig. 3.2 Determination of the open position

is almost impossible to avoid an open position. Additionally, a long validity time of proposals can represent a risk because the portfolio management has to decide if they procure at the moment of presenting the proposal to prevent price risk. However, there is no certainty that the proposal will be accepted. The only way to minimise this risk is to reduce the validity time of proposals. Finally, the variable duration of contracts also represents a risk, as we have explained.

- **Insufficient liquidity in the long term**. If you invest in a generation asset, you have a long open position in the future. In order to avoid risks, you need somebody to offload the generation to. However, it may be difficult to find an interested counterparty for the very long term, e.g. 10 or 15 years in the future.

In Fig. 3.2 we show a graphical determination of the open position, considering all aspects that we have mentioned. We need to compare current sales volumes with current procurement volumes, ideally in real-time, to manage the open position. With this, the portfolio management decides whether to wait to get a better procurement price or not.

Some sales-related parameters can have an influence on the open position: The market expectation of the energy utility drives the direction and size of the open position. As mentioned earlier, if you believe that prices will rise, you should have a long position. This means that you should buy now even before the sales are made and have the energy "stored".

Additionally, every sale success sends a procurement signal by updating the open position—every sale reduces the open position. If it was positive, it will become smaller or negative. If it was already negative, it will become more negative. The procurement strategy can be limit-oriented or event-oriented, but it is dependent on the sales to determine the amount and structure of the required procurement. Therefore, the sales and procurement teams must work together to define an adequate procurement strategy.

Considering what we have explained in the previous section, procurement strategies can be classified as follows:

- **Limit-oriented**: Determination of the volatility-based price limits at the beginning of the month or quarter. If the limit is exceeded, the purchase is activated, and we use downward dynamic limits. This strategy minimises risks and is typically

profitable for falling price trends. An example of this is the stop-loss we will use in our examples in Sect. 3.2.5.
- **Event-oriented**: Procurement decisions are based on market conditions (e.g. monthly prices, technical analysis, etc.) and the market expectations of the portfolio managers. Fundamental market expectations can be created with the help of energy system models, for instance.
- **Extensions**: Apart from standardised products and strategies, more sophisticated approaches can be included. This can be for instance spot market transactions, option trading, trend-following strategies, etc.

This open position-oriented strategy (also known as holistic approach) has several advantages: It is focused on the company's results, is transparent and enables the calculation of key metrics, such as "Value-at-Risk" (VaR), which will be further elaborated in Sect. 4.2.2.

There are also some disadvantages which need to be considered: The "open position approach" has high data requirements and not every company can provide this data in real-time. Additionally, the different departments of the company need to work together, which does not always happen in practice. Therefore, the right interfaces and profit responsibilities should be in place.

Before market liberalisation, an intensive management of the open position was not strictly necessary. Back then, it was common that a comprehensive supply contract was signed once a year between the company and an energy supplier without any risk analysis.[14] For larger companies, however, this has changed since then. Today, they optimise the risk-return expectation by trying to lower procurement costs through active trading. Of course, there are risks associated with such management of a portfolio. To deal with these risks, a high degree of professionalism, significant market knowledge and administrative support are needed. Additionally, there is a lot of regulation in place for energy trading, because it is a high-risk business which may even impact society or the financial system. This regulation obliges compliance with reporting and capital requirements, licences, etc.

Depending on the size of a company, as well as its strategy and key objectives, portfolio management can either use internal resources of a company by developing its own capacities or it can be outsourced to external sources. The latter implies that the company still receives the portfolio management service and benefits from price fluctuations, even if it does not carry out the trading itself. As an example, this could be a good solution for a large industrial consumer who does not consider energy its core business but still faces significant energy costs.

Figure 3.3 shows the different procurement concepts as a function of effort, market opportunities and trading activity. From the all-inclusive contract on the front left, which has low effort but also low market opportunities and low trading activity, we can increase the complexity until we reach the very right back, having an own trading department. This is a fully fledged solution, but there are other intermediate approaches, e.g. having portfolio management without taking too much risk or the

[14] Today, this is still true for small industrial customers, because it is not their core business and they do not want to spend resources in managing an integrated portfolio.

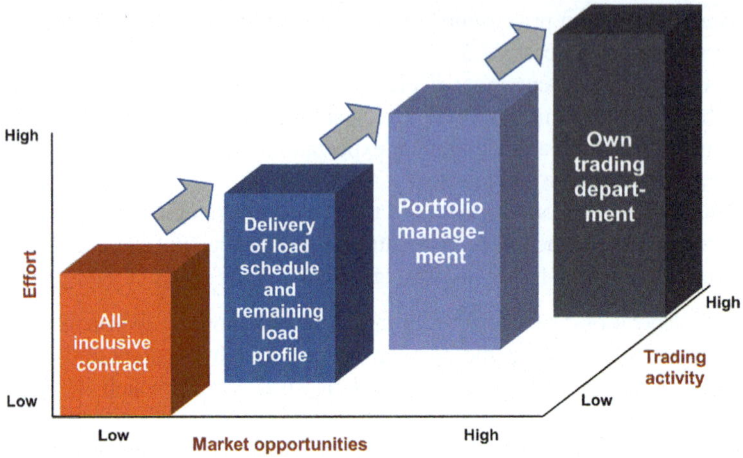

Fig. 3.3 Alternative procurement concepts

delivery of load schedules and remaining load profiles. For instance, instead of having an all-inclusive once-a-year contract, a company could take advantage of fluctuations in forward prices and decide to buy some of its energy quarterly.

3.2.4 *Revenue Generation*

In most companies, the portfolio management process described in this chapter is a profit centre. Hence, in this section we elaborate what factors in the process determine revenues and what can be done to optimise these factors—and the resulting revenues.

We know the utility buys energy on the wholesale market and sells it to retail consumers. The utility's profits must therefore come from selling energy at a higher price than from buying it. Regarding revenues, the relevant question is hence what can be done to lower procurement prices compared to sales prices—or raise sales prices compared to procurement prices. Both are sides of the same coin, but the change in perspective is nonetheless helpful.

Following this line of reasoning, revenues can be increased in three major ways:

1. **Having an excellent sales team**. This means the sales team realises attractive conditions when negotiating contracts with retail customers. Typically, attractive conditions from the perspective of the energy trading company translate into a sales price above current market value of the demand schedule sold to the customer. To achieve this, it is helpful to know the customer well, to develop a brand for the company or its products, to sponsor local events, etc. From the perspective of potential clients, the utility has to deliver a better package than its competitors, since retail customers can buy from anyone in the market. However, if you have an excellent sales team, clients might buy from you even though you

may not have the lowest price compared to the competition. Nonetheless, it is a challenging task to excel in sales activities on a relatively competitive retail market.

2. **Having an excellent procurement team**. This means that the procurement team is able to buy the required energy at low prices. It should be noted that the skills required differ from those of the sales team: the wholesale market price cannot be negotiated. A price on a broker screen is "take it or leave it". Nevertheless, there are ways to optimise the procurement price. The most important is to select the right point in time for the procurement deal: the delivery period is fixed through the schedule, but the energy can essentially be bought any time during the trading period.
3. **Keeping costs down within the utility**. Even once the gross margin of a deal has been determined, i.e. both sales price and procurement price have been fixed, the profit still depends on the associated costs within the company. Efficient processes and a small but effective team in the organisation, for example, increase profits even with a relatively low gross margin.

There are different procurement strategies, depending on the procurement goal and risk specifications. From the trading point of view, there are low-risk and high-risk strategies. The differences are shown in Fig. 3.4.

Low risk means that the municipality has none or hardly any open position. This is called back-to-back procurement. Essentially, when the sales team sells electricity to an industrial customer they report it to the procurement team, and the portfolio management immediately balances the deal by buying the same amount on the wholesale market. However, while this does not involve any price risk, it does not involve any chances either. Thus, the margins must come from the sales strategy and keeping overhead costs low.

If the utility decides that procurement should also contribute to revenues, it will have to take risks because there is no chance without risk. The objective is to choose favourable procurement times and identify low prices. The main goal is to buy at low prices. This can be done even before the customer has bought the electricity and only based on expected sales. Another simple strategy can be that, if portfolio

Low risk:

- No/hardly any open position
- Margins come (primarily) from sales and low overhead cost

Taking on risk:

- Choice of favourable procurement times (identifying "low" prices)
- Speculative portfolios (variety of strategies)
- Requires risk management and risk capital
- Margins should also be realized through procurement below sales price

Fig. 3.4 Procurement strategies regarding risk

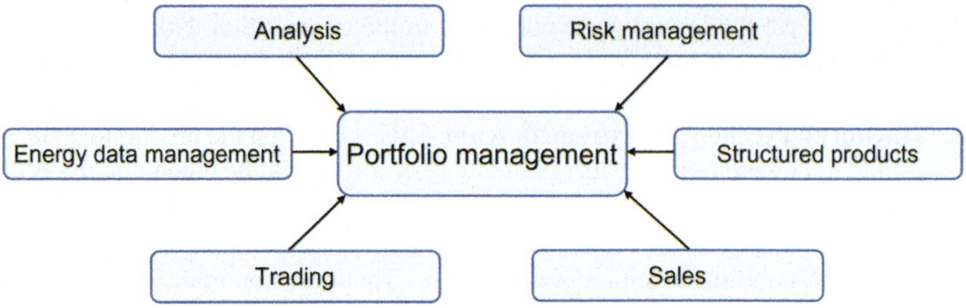

Fig. 3.5 Different aspects involved in portfolio management

management expects a falling price trend, they can leave a negative open position after the sales team has sold to the customer. For instance, they could decide to wait until the next day, and if the electricity is cheaper then, they can either buy and secure a profit or they can decide to wait even longer and hope the price trend continues. Then maybe a month later if the price is even lower, they can buy and secure an even higher profit. However, this implies taking risks, because it is not known in advance if prices really go down or maybe go up instead. Hence, such strategies require risk management and risk capital. There is a variety of possible strategies, and the final idea is always that margins should also be realised through the procurement side.

To achieve this, companies need a high degree of professionalism and market knowledge, as well as great administrative effort. Utilities and large consumers have the option of developing their own capacities, outsourcing this service, or signing cooperation agreements. This usually depends on the size of the company. Large companies prefer to have their own capacities, but for small utilities, this may be too costly. Additionally, there are companies specialised in providing consulting services to municipal utilities which are too small to have their own portfolio management.

When making this decision, utilities must consider that there is a lot of complexity involved in portfolio management. In Fig. 3.5 you can see the main elements that need to be considered. Portfolio management is becoming increasingly prevalent, not only among retail companies but also among generators for the marketing of their generation and purchasing fuels and CO_2 emission certificates. To conclude this section and summarise the previous content so far, Fig. 3.6 illustrates the different phases of the portfolio management process before it will be discussed in detail, including examples in the next Sect. 3.2.5 .

3.2.5 Portfolio Management Process

In the following, we will go through the portfolio management process and discuss two examples. In the examples, we simulate two alternative procurement strategies over time and compare the outcomes. The numbers are based on price curves for

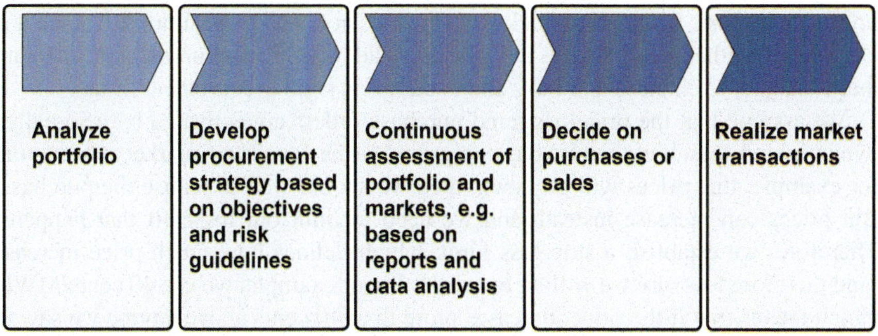

Fig. 3.6 Portfolio management process

a historical delivery year, and we use automatic purchase triggers which are purely based on the price.

This has three advantages and one disadvantage. First, it simplifies the simulation and clarifies the functionality and possibilities of the procurement strategy. Second, the automatic triggers are well suited for an ex-post analysis where it would otherwise be tempting to simply look at the lowest price during the historic period and pick it as an optimal procurement date. This of course is impossible in real-time portfolio management where every decision needs to be taken based on the information available on that day t, i.e. without knowing how the price will develop afterwards. Portfolio management is both challenging and interesting, as we have to make ex-ante decisions without having complete information. Outcomes are different from what is optimal ex-post.[15] Third, a rigid limit strategy purely based on prices avoids irrational short-term decisions and therefore can be part of procurement strategies in practice.

In addition to these advantages, the automatic triggers in the example also have a considerable disadvantage: they are too rigid. In reality, the portfolio manager's expertise or some market developments may influence the decision. The decision could also be based on fundamental market analysis, but that is way more complicated to explain in retrospect. While the market is inherently uncertain, a good portfolio manager with a good fundamental model can still identify trends before they manifest in market prices. This is because markets are mostly but not always correct.[16]

We try to abstract from all of this in the examples and use a simple procurement strategy that can be carried out and evaluated purely based on price realisations. To define a strategy, first we need to know the company's risk attitude, i.e. whether it wants a back-to-back strategy or is willing to take certain risks to make additional

[15] This is like throwing a dice and trying to predict whether it is going to be odd or even. Both have 50% probability, so you pick one and that is the ex-ante decision. Once the dice is thrown, it is pretty easy to answer what was the right choice, but then it is pointless.

[16] For instance, if you knew that the dice from the example above has a probability of 75% (instead of 50%) for an even number, you will put your money on even. You have an expected win, but the outcome may still be an odd number.

profits. In a sense, the back-to-back strategy does not need to be simulated since it is clear that it neither faces chances nor risks. Instead, we use a procurement mechanism that is subject to chances and risks and is based on rigid procurement limits.

We assume that the price-triggered purchase orders are activated by a so-called dynamic stop-loss. If we need to buy energy and we have a certain market expectation, for example, that prices will go down, then we should wait to realise the purchase. But prices can increase instead, and we need to limit our losses if that happens. Therefore, we establish a stop-loss limit, which defines how much price increase (and therefore loss) are we willing to accept. In our example, we use 40 cents/MWh, which means that if the price increases more than 40 cents above a price we saw in the past, we buy.

The dynamic part means that we adjust this limit whenever we see a lower price. Assume that we start our portfolio management today with yesterday's settlement price of 50 €/MWh. This means that the stop-loss limit is initially set to 50.4 €/MWh. If the first observed settlement price today is 49 €/MWh, then we congratulate ourselves because we have earned 1 €/MWh and we made the right decision by not buying at 50 €/MWh. Now we hope that the price will continue to fall. In this case, we did not reach our limit, which on the first day was set at 50.4 €/MWh.

If the price had gone up instead, we would have made a loss. As soon as it reaches the stop-loss limit at 50.4 €/MWh, the procurement would have been triggered and we would have bought. We probably would not have bought exactly at 50.4 €/MWh, but maybe around 50.5 €/MWh, because the procurement process takes some time and may involve transaction costs. Consequently, it is not guaranteed that the limit will be the purchase price. The idea, however, is to avoid an irrational decision like waiting and hoping that the price will go down again with things getting worse and worse.

Now let us go back to the original assumption where the price went down to 49 €/MWh. In this case the dynamic characteristic means that we adjust the limit from 50.4 €/MWh to 49.4 €/MWh, since 49 €/MWh is the lowest settlement price seen so far. Next, we assume that the next day we have a small price increase to 49.2 €/MWh. In this case, we must not adjust the limit, which stays at 49.4 €/MWh. Since it was not reached the procurement goes on.

On the next day, we assume that the price goes down to 48 €/MWh, and we congratulate ourselves again because we have now earned 2 €/MWh in total. The new reference price is now 48 €/MWh, and the new limit is 48.4 €/MWh. If the next day the price increases to 48.4 €/MWh and we buy at, say, 48.45 €/MWh, we did a very good job because we saved around 3% (1.55 €/MWh from the original price of 50 €/MWh), which can be a significant amount of money. We can say that in this case the reference price is stable upwards and dynamically downwards.

The dynamic stop-loss can be used to procure standard products and schedules. Standard products have already been defined as base and peak contracts. We define a *schedule* as an energy time series. This can be for instance the 96 periods of 15 min of the following day, or the 8760 hours of a year. Typically, the values in a schedule

Table 3.1 Detail of the structured procurement strategy

Date	Schedule	P_{max} (MW)	Energy (GWh)
Quarter IV 01/10/y−2—31/12/y−2	1	35.25	133.50
Quarter I 01/01/y−1—31/03/y−1	2	35.25	133.50
Quarter II 01/04/y−1—30/06/y−1	3	35.25	133.50
Quarter III 01/07/y−1—30/09/y−1	4	35.25	133.50
Total		141.00	534.00

vary, e.g. following a demand profile.[17] Trading a schedule thus requires an exchange of the time series data in question, e.g. in CSV format.

However, in real-life portfolio management, strategies are more complex, and a permanent evaluation of activities and an iterative improvement process of the procurement system should also be implemented. Now, let us take a look at the two empirical examples.

3.2.5.1 Example—Scheduled Procurement

Assume a utility wants to procure a total amount of electricity amounting to 534 GWh in one year, and the maximal load is 141 MW during that period. The procurement strategy in this empirical example is simple: instead of buying everything at once, we want to procure four identical load profiles or schedules. In order to get a solidly distributed risk-averse procurement price, each of the four schedules shall be procured within one of four consecutive quarters. We know that the total amount of energy is 534 GWh, and the maximal load is 141 MW, so we take four identical slices out of this profile. We consider the stop-loss related to the base product with a spread of 0.4 €/MWh and the reference price corresponds to the closing price on the exchange on the first day of each quarter.

In Table 3.1 we see in detail how the structured procurement is established. As you can see, two years before delivery, in the fourth quarter (between the 1st of October and the 31st of December) we want to buy one of these schedules with a total energy amount of 133.5 GWh and a maximal power of 35.25 MW.

In the first quarter of the year before delivery, we want to buy the second schedule and so on in the remaining two quarters. With this, we fulfil the requirement of completing the procurement process in the third quarter of the year before delivery and thus have the fourth quarter to prepare the delivery plan.

[17] Note however that standard product delivery can also be interpreted as a schedule. For example, a 5 MW base contract for the front year specifies a schedule for the 8760 hours of that year, i.e. an energy delivery of 5 megawatts in each hour of the year.

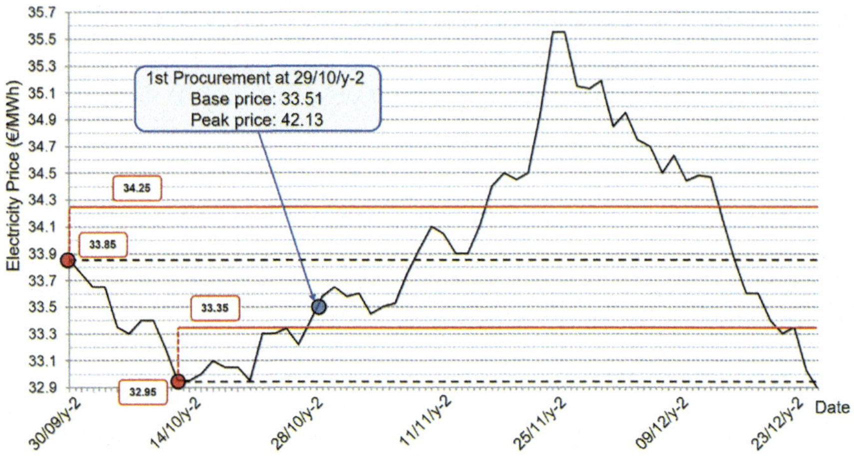

Fig. 3.7 First procurement in the fourth quarter in year $\boldsymbol{y}-2$

As we said, the first procurement occurs during the fourth quarter two years before delivery. Figure 3.7 shows the development of the base load price in this quarter for the delivery year *y*. As you can see in Fig. 3.7, the opening price was 33.85 €/MWh, and in consequence, our first limit was 34.25 €/MWh. Then, during the first days of the quarter, the price fell, which is good news for us in terms of procurement, because then buying on the first day would have been unfavourable. Our strategy of waiting was good. Afterwards, the price dropped further to a minimum price of 32.95 €/MWh, which lead to the stop-loss going down to 33.35 €/MWh. After that, around the 14th of October, the price went up again, and on the 29th our dynamic stop-loss limit was reached, and we bought.

The price chart in Fig. 3.7 depicts the base price, and we use it as a reference because this is a standard product that is liquidly traded and thus prices are reported. However, we need to buy a schedule and not the baseload product, consisting of one fourth of the aggregated consumer profile. We have already discussed the procurement of such a non-standardised product. Depending on the shape of the profile the price can be close to the base prise, to the peak price, or way above it. Nevertheless, it is always highly correlated with the two, which is why we can use the base price as in indicator for our trading strategy in this example.

As shown in Fig. 3.7, the first procurement was executed on the 29th of October with a base price of 33.51 €/MWh and a peak price of 42.13 €/MWh.[18] The schedule's price, which is this example lies between the two, will be reported later. Since we now already bought the schedule due for this quarter, there is no further trading in November and December.

Sometimes it is convenient to benchmark the success of your procurement to the average price of your procurement. In this case this is the average price during the

[18] Keep in mind that the purchase cannot be executed immediately after the stop-loss limit is exceeded.

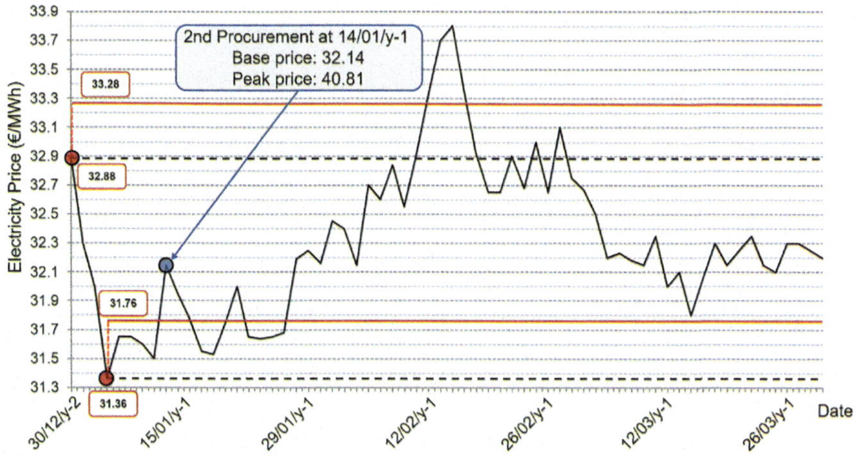

Fig. 3.8 Second procurement in the first quarter in year ***y***−1

quarter. Once you have bought the amount for one quarter, you hope that prices increase during the rest of the quarter, because that will increase the average price. In this example, the results look good. In retrospect, it is easy to see that the ideal moment to buy was on the 14th of October or by the end of the quarter. However, we did not know that beforehand and the stop-loss order still lead to a good result.

We follow the same procedure strategy for the next three quarters. In Fig. 3.8 we can see the chart for the second quarter, with an opening price of 32.88 €/MWh and a starting limit of 33.28 €/MWh. The price fell to 31.36 €/MWh and the respective stop-loss limit was dynamically adjusted to 31.76 €/MWh. On the 14th of January, the price went above this limit, which lead to a purchase of the second slice at a base price of 32.14 €/MWh and a peak price of 40.81 €/MWh.

We procured the third part of our schedule in the second quarter of the year before delivery. The chart is shown in Fig. 3.9. The purchase was performed on the 3rd of June. As you can see, even in retrospect you could say that it was a good procurement.[19]

Finally, the remaining amount corresponding to the procurement of the fourth schedule occurs in the third quarter of year $y-1$. This is shown in Fig. 3.10. We do this in July and then we have procured the total amount. You can see that in retrospect this fourth procurement was not optimal, because the price dropped significantly in August and September. However, this was not known in advance and adhering to our strict stop-loss strategy prevented losses from a possible rising price.

In Table 3.2 you can see a summary with the date of the procurements, the base price, the peak price and the schedule price, which is somewhere in between the peak and base prices. The mixed price is the average of the schedule price for the four

---

[19] We eventually bought at a base price of 31.62 €/MWh, which is 1.7% below the price at the beginning of this quarter and (through visual assessment) also seems to be below the average price during the quarter.

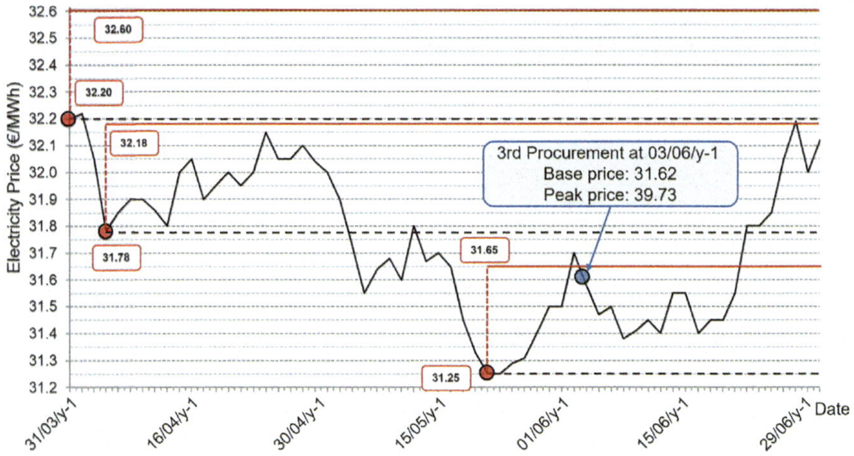

Fig. 3.9 Third procurement in the second quarter in year $y-1$

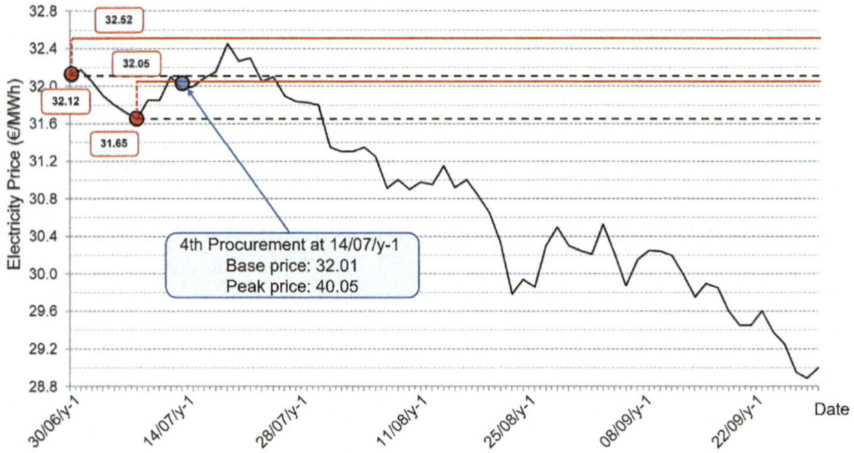

Fig. 3.10 Fourth procurement in the third quarter in year $y-1$

procurements. As this value is the key result to benchmark the portfolio management strategy, it is underlined in the table. Historically, a mixed price of around 34 €/MWh is comparably low for European countries. Multiplying this mixed price with the total amount of 534 GWh leads to the total procurement cost in euros.[20] As you can see, it is a huge amount of money, even though this is only for the household customers of an average size municipality and mixed prices have mostly been above 34 €/MWh in Europe.

[20] If you calculate this, please bear in mind that there may be rounding errors.

Table 3.2 Procurement result

Date	Base price [€/MWh]	Peak price [€/MWh]	Schedule price [€/MWh]	Mixed price [€/MWh]	Total price [€]
29/10/y−2	33.51	42.13	35.32	34.08	18,196,370.00
14/01/y−1	32.14	40.81	33.96		
03/06/y−1	31.62	39.73	33.32		
14/07/y−1	32.01	40.05	33.70		

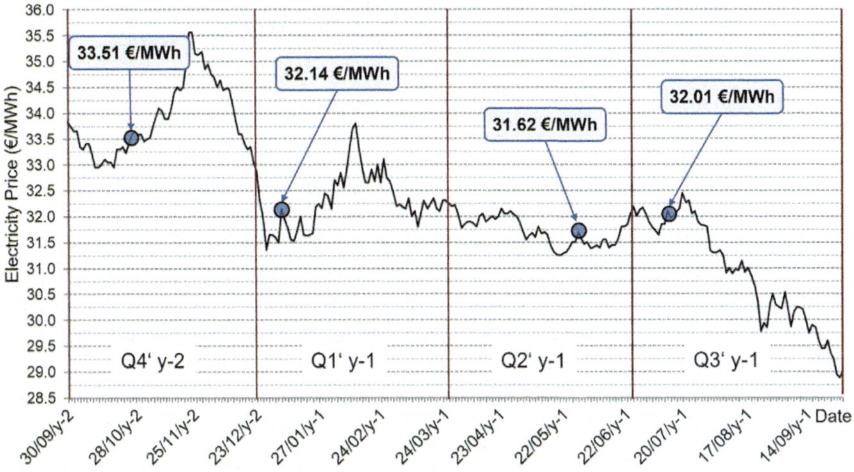

Fig. 3.11 Graphical view of the four procurements

If you assume that with a good procurement strategy you can save 3% of the total cost, that is enough to finance a portfolio manager and e.g. quantitative support from analysts. Three percent of the volumes involved here is significant, and that is why there is usually so much manpower involved in this business.

Figure 3.11 summarises the four procurements graphically. You can see that the first three look good but on the fourth we stopped too early. In the last quarter, it would have been better to wait longer. However, it is always easy to say that in retrospect.

3.2.5.2 Example—Standard Products

In this example, for reasons of comparability, customer demand is again assumed to be 534 GWh with the same maximum load of 141 MW. However, this time the procurement process is performed in a slightly more realistic approach. This time, we want to be more active on the market and buy most of the energy with the standard products of baseload and peak load.

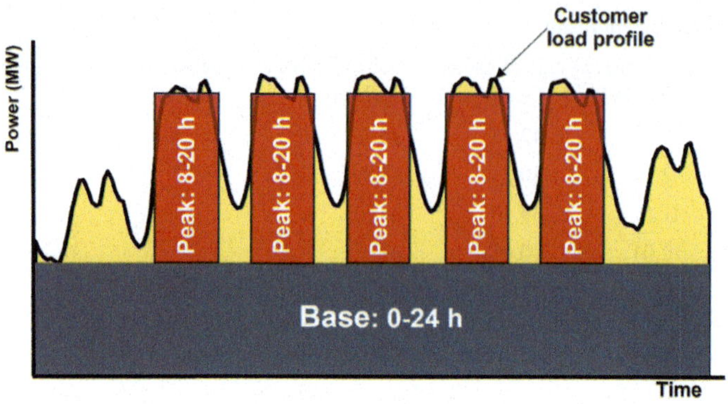

Fig. 3.12 Customer load profile and long-term electricity products

In order to do that, we first need to analyse our schedule and determine how it can be replicated using base and peak products. Assume we see that demand is never below 20 MW throughout the year and therefore we decide that we want to buy 20 MW of baseload. Of these 20 MW, we want to procure 16 MW quarterly using the limit system and the remaining 4 MW on the spot market. Additionally, we procure 28 MW of peak load, of which 24 MW are procured using a trading approach and 4 MW on the spot market.

Before we analyse the remaining schedules in detail, we explain how such a split of a schedule into standard products and a remaining schedule can be done in general. First, remember that only two products have prevailed in mid- and long-term electricity market: base and peak. With these two products, the goals of "good liquidity" (by concentrating on few products) and "good mapping of customer load profiles" are combined in a compromise. This is because we can use base and peak products to replicate the demand curve of a customer. In other words: the two standard products can be used to find the best possible approximation.

This is shown in more detail in Fig. 3.12. The purpose is to approximate the black line, which is the hourly client-specific profile. Here, baseload (shown in dark blue) corresponds to the minimum load of the customer.[21] Then, on top of the baseload product, there is the peak contract from 8:00 to 20:00 (shown in red). What remains to be traded is the yellow part, which is already significantly smaller than the total demand of the customer. This means that using these two standard products to approximate the customer load profile leads to less electricity that needs to be traded in the short-term market. Hence, this gives the advantage of managing risk more easily.

Assume for instance, that the portfolio management of the municipal utility sold the energy according to the customer's hourly profile today without buying it back. This would immediately lead to great financial risk. The reason for that is that the price on the wholesale market may increase or decrease. Since the municipal utility must

[21] The baseload product acquired does not necessarily have to correspond to the customer's minimum demand. In reality, it is often better to buy more baseload and resell the left-over energy back to the market during certain hours with very low demand.

get the electricity from somewhere sooner or later, they can benefit from decreasing prices but suffer from increasing market prices. So, there is a risk to lose money but also an opportunity to make profits, but the coin can flip to any side. If the price decreases on the wholesale market, there is a chance to buy the electricity at a lower price, but there is also the risk that the price will rise, and the utility has to pay more eventually. This is referred to as price risk and will be further elaborated in Sect. 4.2.

If they think the current price is good and they decide to buy on the wholesale market to eliminate price risk, they can buy the dark blue and the red blocks in Fig. 3.12 and thereby significantly reduce price risk. This is the so-called base-peak equivalent which can be used to approximate any given load profile. Doing that will reduce the base and peak price risk dramatically, because the price risk now only concerns the remaining yellow area in Fig. 3.12. Nevertheless, there is still some hourly price risk.

There are some small differences with natural gas. Gas is economically and technically easier to store than electricity, even in large quantities. In addition, natural gas is compressible, so that injection and withdrawal do not always have to coincide. Natural gas pipelines can be considered as short-term buffers. However, the fundamental premise that storage is costly is still valid, but in the very short term, it is slightly easier to balance. Therefore, the standard products for gas in the futures market are a bit different to those in electricity markets. For instance, there only exists a base product in the case of natural gas and no peak product for the period from 8:00 to 20:00. Apart from this, the content of this chapter also applies to natural gas. It is still required to approximate the profile of clients with the available products. However, it is not necessary to use an hourly profile as a basis.

Forwards and futures can be traded on many marketplaces in Europe. We already mentioned EPEX SPOT for day-ahead trades in several European regions as well as the European Energy Exchange for futures trading, and the Nord Pool as a competitor to the EEX and EPEX SPOT, but there are many more.

The introduction of energy markets in Europe was driven by the initiative of the European Union. As a result, in almost every European country, there are now marketplaces for electricity, natural gas, European emissions allowances for CO_2 and several other commodities. Similar markets also exist in many other liberalised countries, which illustrates the relevance of the issues addressed in this book.

Coming back to our example, after deducting the base and peak products, we get the remaining structure of 270.62 GWh, which we procure using schedules. The 270.62 GWh will be divided into two equal blocks of 135.31 GWh each. The first of these blocks is procured at "historically low prices". We will show later how this procurement strategy works. The second block is procured in the third quarter, to adapt to changes in the customer structure. With this block, the idea is to wait until we know, for instance, how many household customers have changed their supplier. Therefore, we want to procure it relatively late. Finally, there is a remaining quarter-hourly structure that needs to be purchased on the spot market.

The structuring result for an example day is shown in Fig. 3.13. There is the base share (dark blue) and the peak share (red), which we can both buy on the OTC market or on the exchange. The two schedules (yellow) are purchased via tenders.

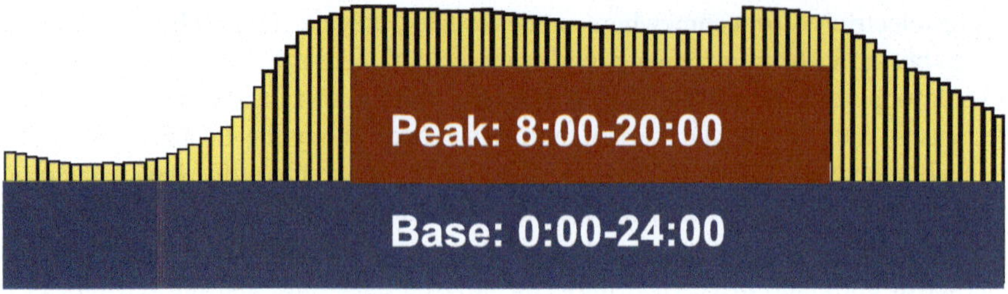

Fig. 3.13 Structuring the load forecast into tradable standard products and remaining load

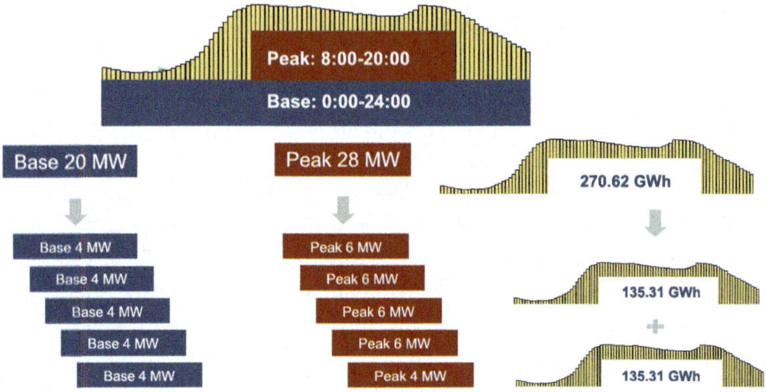

Fig. 3.14 Blocks of electricity to be procured

In Fig. 3.14 we find a visualisation of what we need to procure. This is five blocks of 4 MW baseload (four procured quarterly and the last one on the spot market), four blocks of 6 MW and one of 4 MW peak and the two remaining schedules of 135.31 GWh each. We know now what we want to buy, and now we will see how to buy it.

For the baseload, we need to buy 4 MW each quarter using the limit system that we saw in the previous example. This has essentially the same result as in that example because there we used the base price as a reference for the dynamic stop-loss limit to determine when to buy. As you can see in Fig. 3.15, the procurements are triggered exactly at the same moments as in the previous example, but the difference here is that the base price is really the purchase price.

Now we move to the peak load blocks. Here we present a different trading approach from the toolbox of possibilities: We buy when the daily chart breaks through the six-week moving average from bottom to top. The underlying assumption of this strategy is that the market moves in trends and if the daily price increase is significant and passes certain thresholds such as the moving average, that indicates a sustained rising trend, and we should buy before the price increases even further.[22] We use this

[22] Please not that by applying this strategy, we do not necessarily agree with its assumptions.

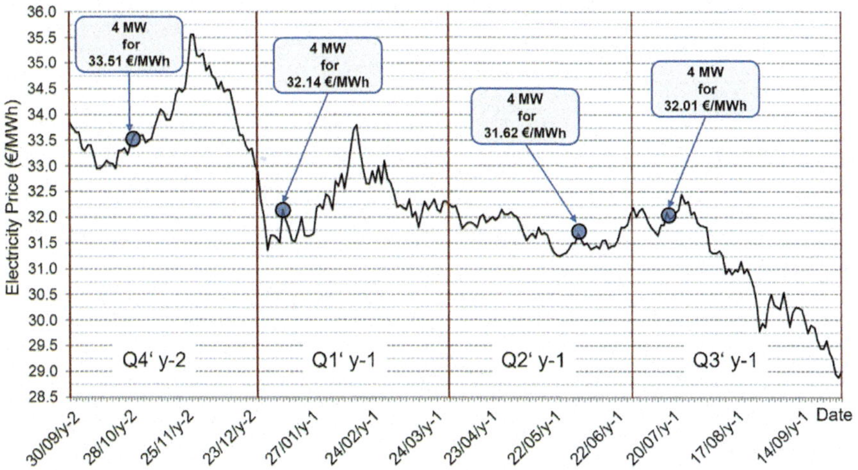

Fig. 3.15 Procurement of the baseload blocks

approach just to show you that there are many strategies used by traders, and you must decide which strategies you trust. Furthermore, we would like to repeat that due to the ex-post nature of this example, we only included automated strategies where predefined rules lead to clear decisions based on the time series of prices alone.

In Fig. 3.16, you can see that we perform our first procurement on the 7th of November. In the chart, the blue dashed line is the six-week moving average and the black line is the daily price. The blue dot indicates the time when the daily price passes the moving average in an upward direction, and the price at the intersection point is shown in the blue box. If you look at the graph, you may say that it was a reasonable decision, because we had a lasting increasing trend afterwards.[23]

The second procurement occurs on the 4th of February, the third procurement on the 22nd of June,[24] and in the last procurement, the price was consistently below the moving average, so we just bought at the end of the quarter. These procurements are shown in Figs. 3.17, 3.18 and 3.19. Again, the purpose of this is just to show you a different procurement strategy and how it is used.

Finally, we have the two remaining load profiles in the two schedules of 135.31 GWh each. The first schedule will be procured using "historically low prices". We say that prices are historically low when the two following conditions are met simultaneously:

- The current price is below the moving half-year average.
- The moving 6-week average begins to rise and breaks through the moving half-year average from below.

[23] This, however, is again an ex-post perspective.

[24] Here, we were not able to buy at the market price when the moving average was broken, but at a price slightly above it.

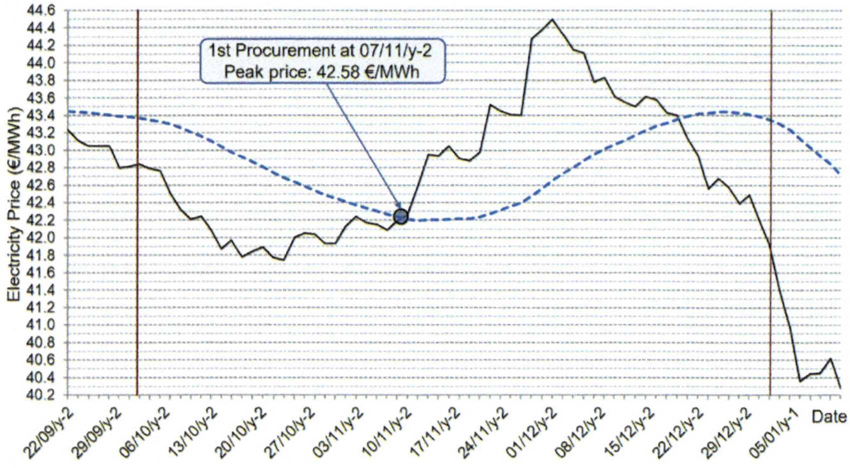

Fig. 3.16 Procurement of the first peak load block

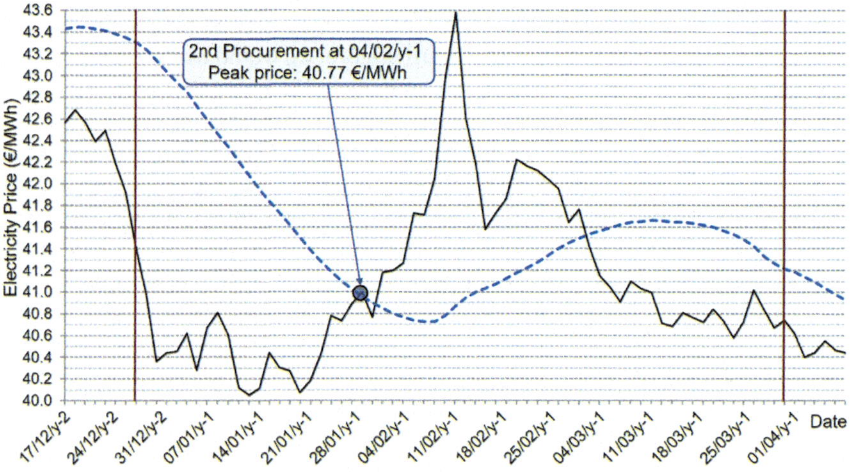

Fig. 3.17 Procurement of the second peak load block

Figure 3.20 shows how this works. The red line is the moving six-week average, and the light blue line is the moving half-year average. In this figure, you can see when the procurement of the first schedule happens at "historically low prices".[25] The procurement of the second schedule is also shown.[26] Both are procured in the

[25] Read again the conditions, if necessary. This will show why the procurement occurs at this point.

[26] Remember that the procurement of the second schedule was planned to adapt to changes in the customer structure and therefore happened relatively late.

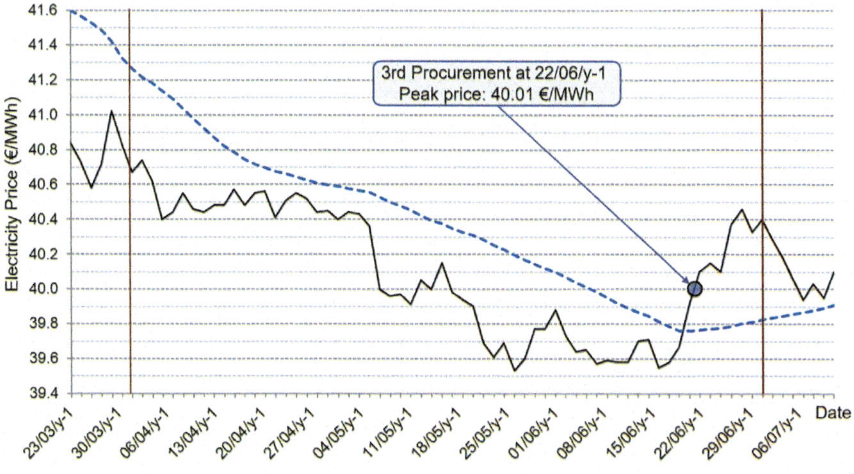

Fig. 3.18 Procurement of the third peak load block

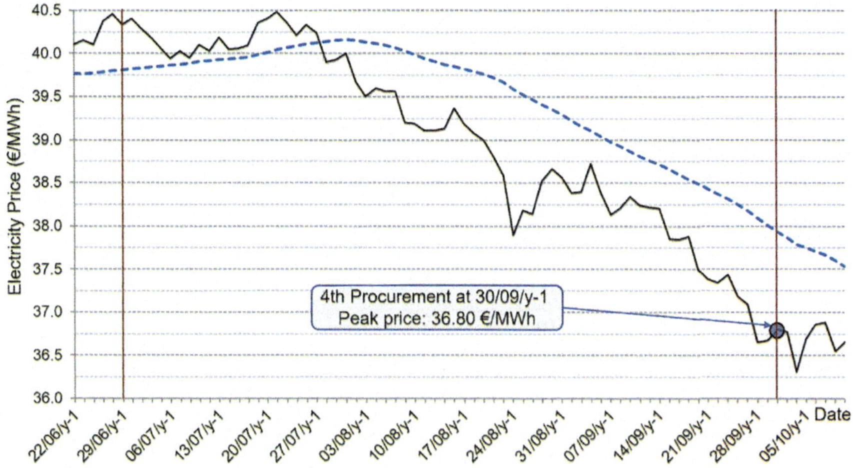

Fig. 3.19 Procurement of the fourth peak load block

third quarter: the first one on the 28th of July and the second one on the 28th of August.

In Table 3.3 we have the result of this procurement strategy. We see base, peak and schedule prices, according to what we have previously seen, and also the spot market procurements. We see that in total we bought 534 GWh, we paid roughly €17.88 million and the average price was 33.49 €/MWh.

Figure 3.21 summarises the procurement process timelines for the procurement strategies of examples 1 and 2. Here we show when each of the blocks was purchased

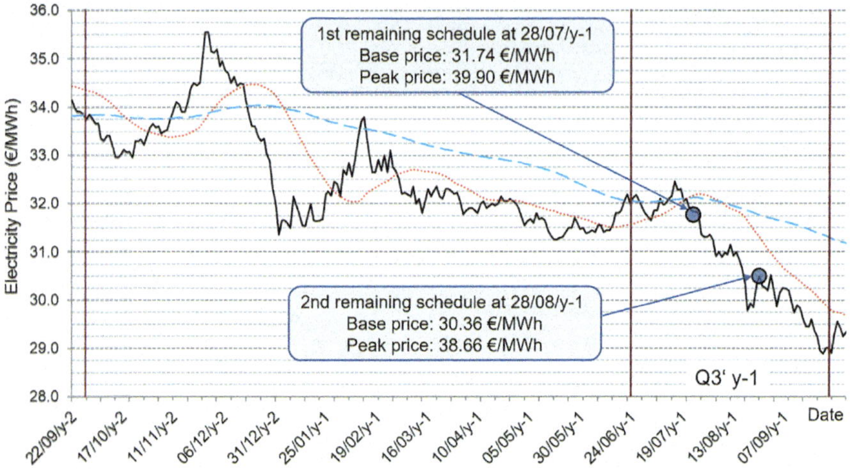

Fig. 3.20 Procurement of the two remaining schedules

Table 3.3 Procurement results using the second strategy

Date	Base price [€/MWh]	Peak price [€/MWh]	Schedule price [€/MWh]	Quantities [MWh]	Individual prices [€]
29/10/y−2	33.51			35,136	1,177,407
07/11/y−2		42.58		18,792	800,163
14/01/y−1	32.14			35,136	1,129,271
04/02/y−1		40.77		18,792	766,150
03/06/y−1	31.62			35,136	1,111,000
22/06/y−1		40.10		18,792	753,559
14/07/y−1	32.01			35,136	1,124,703
28/07/y−1			33.46	135,312	4,527,742
28/08/y−1			32.10	135,312	4,343,921
30/09/y−1		36.80		18,792	691,546
01/01/y−31/12/y spot market 4 MW Base			28.98	35,136	1,018,329
01/01/y−31/12/y spot market 4 MW Peak			35.24	12,528	441,521
Totals				534,000	17,885,313
Ø total price					33.49 €/MWh

and the percentage of the total demand, we also include the amounts purchased on the spot market.

In Table 3.4, the resulting prices of the two strategies are compared with the average price during the procurement period as a benchmark. As you can see, strategy

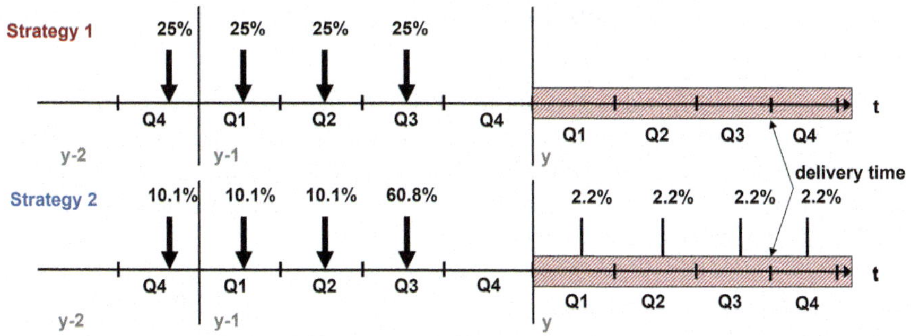

Fig. 3.21 Comparison of strategies

Table 3.4 Comparison of the results of each strategy

	Prices achieved [€]
Benchmark [Ø OTC: Base 32.18; Peak 40.71]	33.97
Strategy 1	34.08
Strategy 2	33.49

one is slightly more expensive than the average, and this is not good news because we could have simply bought the average, which is a product offered by many trading houses.[27] Since you can buy the average without the need for portfolio management, being more expensive than that is bad news. On the other hand, our second strategy had positive results, because we were 50 cents cheaper than the average.

These two procurement strategies work with automatic order signals. We replicated them and simulated their performance using historical prices. Hence, the outcome cannot be generalised, and we would recommend not to use such automated signals in practice—at least not exclusively. We rather suggest analysing the markets fundamentally and we would also consider the output of energy system models in order to determine price developments, such as the price for the front-year contract. Of course, models are always a simplification of reality. But if, for instance, there is a strong deviation between the fundamental model and the market price, and the model persistently finds the front-year base price should be at 30 €/MWh whereas the market trades it at 50 €/MWh, the market price may be too high, or at least that it cannot be explained fundamentally. The remaining question is whether the (assumed) bubble will burst before delivery and thus the spot price will be cheaper and make trading decisions based on that.

But as we mentioned before, there is a variety of trading strategies, and other people with different backgrounds may prefer different strategies. In the end, the importance is to be successful and make money for the company.

[27] You can always replicate the average by buying a very small piece of the total volume each trading day. If you have 534 GWh which you want to purchase over 200 trading days, then you buy $534/200$ GWh $= 2.67$ GWh per day.

3.3 Energy Utility with a Generation Focus

In this section, we will move the focus towards generation assets. The company now generates (or more correctly *converts*) its own energy using generation plants and then sells this energy on the wholesale markets. Such activities are performed by different types of companies, for instance operators of conventional power plants, wind farms, large-scale solar projects, or gas producers.

From an economic perspective, the process of producing and selling is always similar, independent of the generation type. We will focus on thermal power plants selling their production on the wholesale electricity markets. However, the "lessons learned" in this section are also valid for gas producers. This sales process requires in-depth knowledge of the markets and the various products used for the sale of self-generated electricity and products derived from it. Apart from the "general" products we introduced in Chap. 2, there are other products such as spreads, swaps and options, which will be presented in this section because they are closely related to the flexibility of generation assets. In the last sections we will discuss how flexibility can be modelled and marketed.

3.3.1 Spreads and Profitability of Power Plants

Before selling power from a generation unit, it is first necessary to understand what determines the profitability of a power plant. There are several economic factors that can be considered to do so. For example, the profitability of a power plant can change in line with economic developments, political decisions or technological changes. Therefore, to find an answer to what actually drives its profitability, we need to simplify as much as possible and think about what a thermal power plant typically does: It burns fuel to generate electricity. In economic terms, this means that the power plant operator pays for fuel (short position) and receives income from the sale of the electricity (long position). This relation can be formalised with the so-called **spark spread** product.

A spark spread is a non-standardised, cross-commodity product on the energy markets. Its payout is the difference between the electricity price and a fuel price. More precisely, in a spark spread, the fuel is natural gas, and the payout π at hour h is

$$\pi_h = S_h - G_h/\eta, \tag{3.3}$$

where S_h and G_h are electricity and fuel prices at hour h, respectively and η is the efficiency of the power plant, i.e. the share of input energy that is converted into output energy. This is what drives the short-term profit contribution of the power plant.

However, fuel is not the only commodity that is "used" when operating a gas-fired power plant. In Europe and many other countries worldwide, there is a penalty for emitting carbon dioxide. This penalty can be a tax or the requirement to purchase certificates for each tonne of CO_2 that is emitted. In any case, there is a price for emitting one tonne of CO_2 in these jurisdictions. This price must be considered when determining the power plant's profitability. For that purpose, there is also the so-called **clean spark spread**. It takes into account both fuel costs and the costs for CO_2 emissions. The payout of the clean spark spread at hour h is

$$\pi_h = S_h - \left(G_h + E_{CO_2} \cdot C_h\right)/\eta, \tag{3.4}$$

where C_h is the price for CO_2 emission certificates at hour h, and E_{CO_2} is the emission factor, i.e. the tonnes of CO_2 emissions that are emitted by burning one MWh natural gas. The rest of the terms of Eq. (3.4) remain the same as in Eq. (3.3). There may be other factors influencing the short-term profitability, such as cost for lubricants. However, these are difficult to standardise and can often be neglected.

The clean spark spread involves natural gas as fuel. However, there are other fuels that a typically used to generate electricity, particularly hard coal. Therefore, there is also the **clean dark spread,** where the fuel is hard coal and the **clean brown spread** when the fuel is lignite. Clean spreads and spreads in general are traded as financial products, mostly over-the-counter. The next example shows the calculation of a (clean) spark spread.

Example: Calculation of a spark spread, and a clean spark spread.

- Electricity price: $S_h = 48.44\ €/MWh_{el}$
- Gas price: $G_h = 18.71\ €/MWh_{th}$
- Efficiency 55%, i.e. MWh_{el}/MWh_{th}

The spark spread is $\pi_h = 48.44\frac{€}{MWh_{el}} - 18.71/0.55\frac{€}{MWh_{el}} = 14.39\frac{€}{MWh_{el}}$

- CO_2 price: $C_h = 41.76€/t$
- $E_{CO_2} = 0.2011\ t/MWh_{th}$

The clean spark spread is $\pi_h = 48.44\frac{€}{MWh_{el}} - \left(18.71\frac{€}{MWh_{th}} + 0.2011\frac{t}{MWh_{th}} \cdot 41.76\frac{€}{t}\right)/0.55 = -0.90\frac{€}{MWh_{el}}$.

The results of the example are shown schematically in Fig. 3.22. With this calculation, we can estimate whether a power plant makes profits or losses. If the power plant does not have to pay for its CO_2 emissions, it makes a profit contribution of 14.39 €/MWh. However, if there is a price for CO_2 emissions, the clean spark spread has to be used, which in this example is negative, i.e. the power plant is making losses.

As the clean spark spread is negative, the power plant operator will decide not to operate the power plant at this hour. This is because if the power plant generates electricity, the costs of electricity generation would exceed the income from the sale of electricity. The power plant therefore makes a loss.

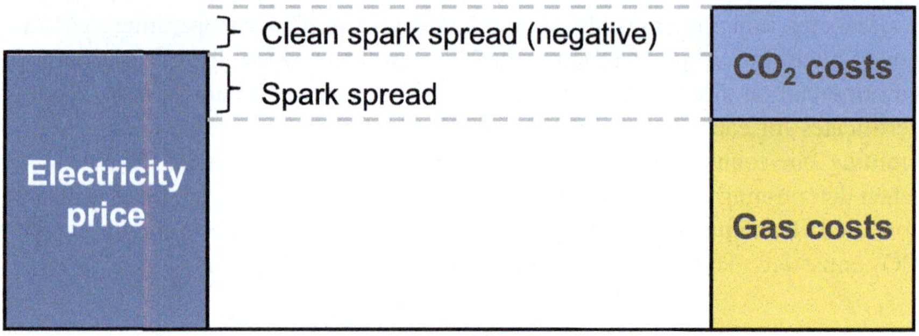

Fig. 3.22 Spark spread and clean spark spread compared to electricity price

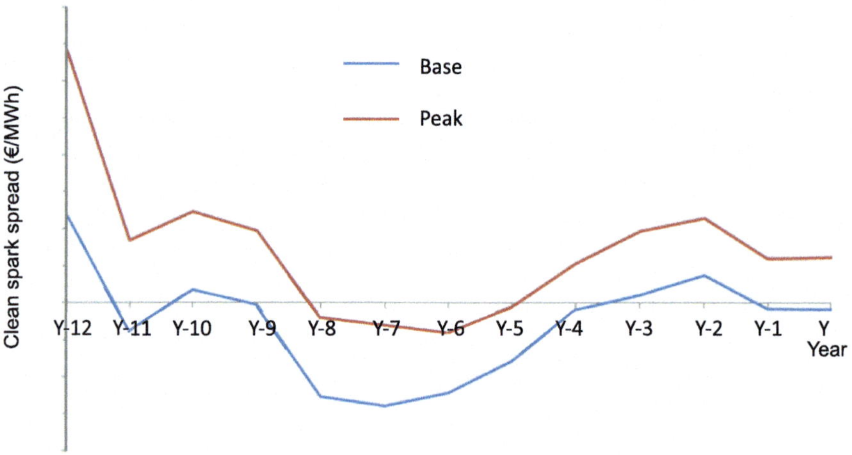

Fig. 3.23 Historical development of the clean spark spread in Germany

Figure 3.23 shows an example of the price evolution of a historical annual clean spark spread.[28] It is calculated for both base and peak prices. In the former case this means that this is the average clean spark spread of all hours of the year. In the latter case it is the average clean spark spread of all hours of the year that lie in the peak period, i.e. all workdays between 08:00 and 20:00. The figure shows that that during the base period (all hours of each year), the clean spark spread for a gas-fired power plant was mostly negative, except for a couple of years. In other words: if a power plant was in operation in all hours of the year, it made a loss. This is in line with observations of the actual operation of such power plants. In Europe, gas-fired power plants usually operate only in the most expensive hours, with operation hours ranging from a few hundred to a few thousand hours per year.

[28] The parameter η is assumed to be 50%, which is a reasonable assumption for gas-fired power plants, and $E_{CO2} = 0.2011$.

For this reason, the figure also shows the peak clean spark spread. Since electricity prices during the day on workdays are usually higher than average, the peak clean spark spread is also higher and, in some years, significantly positive. However, this does not mean that this plant will generate electricity every day from 08:00 to 20:00. The peak clean spark spread is simply a proxy for the profitability of gas-fired power plants in the most expensive hours. Still, in some hours the prices will be low, i.e. the clean spark spread will be negative and, in such cases, the plant's operator will decide not to generate electricity. However, there are additional costs and constraints for a power plant's operation, such as start-up or maintenance costs, which are not considered in the clean spark spread, but play an important role in real-time operations. The clean spark spread is therefore a simplification that provides an indication of the general profitability of a gas-fired power plant.

Figure 3.23 shows the average profitability in base and peak hours over a couple of historic years. However, what is driving the inherent value of a power plant are the expectations about the future. If for example a company needs to assess whether a power plant should continue operating or whether an investment in a new power plant is more favourable, it is reasonable to look at the past to get a rough idea of how the market has behaved. However, what is way more important for the investment decision are the expected electricity, CO_2 emissions certificates and gas prices for the future, i.e. the future (clean) spark spread. For instance, in order to calculate the base clean spark spread for next year, the year-ahead future prices for baseload electricity, natural gas and CO_2 certificates need to be used in the calculation. The formula given in Eq. (3.4) remains the same.

In competitive markets, market forces tend to correct these spreads because no one will invest in new generation assets that are not profitable. The clean dark/spark spread not only needs to be positive but also high enough to cover the additional costs of running a power plant, i.e. all fixed costs. If this is not the case, no investor will be interested in building new plants. Once clean dark/spark spreads have been low for a while, existing (old) power plants start to leave the market. Demand may increase. Consequently, forward prices for electricity will rise, which increases prospective new plants' profitability, and investors will be interested in building power plants again.

To return to the question posed at the beginning of this section: As we have learnt, expectations for the future are the true driver of a generation asset's value. Market participants need to develop their own expectations about the future price development and expected profitability of a power plant in the future. A good parameter to estimate this profitability is the (clean) spark spread (gas-fired power plants), the (clean) dark spread (coal-fired power plants) or the (clean) brown spread (lignite-fired power plants), parameterised with expected future fuel, electricity, and CO_2 emission price data.

3.3.2 *Swaps*

In the previous section, we have learned how spreads reflect the profitability of power plants. These spreads can be traded as financial products. In Sect. 2.5 we have also learned how forwards and futures work. In this and the following section on options, we build the bridge from energy generation to energy trading and portfolio management. Spark spreads are an introduction to financial instruments and financial trading products. They are a simplification of the complex machinery behind a real power plant, focusing only on the costs (incoming gas) and revenues (outgoing electricity). Swaps, on the other hand, are an application of forwards, which are commonly used in energy markets. They may, but do not have to, include a dark or spark spread formula.

Essentially, an energy utility with a generation asset can use a swap to transform an uncertain future revenue stream into an ex-ante known, constant revenue stream. The utility still markets the asset, i.e. buys fuel and sells electricity, but the swap financially compensates the difference to an agreed reference revenue.

We start the more detailed analysis of swaps with an introduction and show later how swaps are used in energy markets to market electricity generated in power plants and manage the associated risks. We define a swap as follows: *A Swap is an agreement to exchange cash flows at specified future times according to a prearranged formula.*

This definition is fairly broad, but it certainly implies that swaps are settled financially. Swaps are non-standardised products and thus traded only OTC. The rules for the exchange of the money within a swap are defined and agreed upon when the contract is concluded. They often comprise future prices, especially of commodities, and interest rates. In principle, a swap can contain any future price that can be described with a formula. In energy swaps, variable payments are often exchanged for fixed payments. While the content of a swap in principle can often be replicated with other financial contracts (e.g. with numerous futures), the use of a swap can be more straightforward and save transaction costs.

Before we discuss the application of a swaps to secure a fixed future cash flow from a (gas-fired) power plant, we need to understand how swaps are constructed and what they are typically used for in practice. After that we will discuss how they can be applied to energy markets.

3.3.2.1 Swap as Combination of 2 Physical Trades

This section presents the underlying concept of swaps as the combination of two physical trades: the purchase/sale of physical energy at a fixed price with simultaneous sale/purchase of the same energy with the same maturity and quantity and a variable (or index) price. Consider the following example in Fig. 3.24.

In this example, the marketer (an actor interested in exchanging a financial payment stream with a swap) purchases natural gas from a counterparty and pays a fixed price of 20 €/MWh (left part). At the same time, the marketer sells that gas

Fig. 3.24 Swap as a combination of physical trades 1

Fig. 3.25 Swap as a collapse of physical trades 2

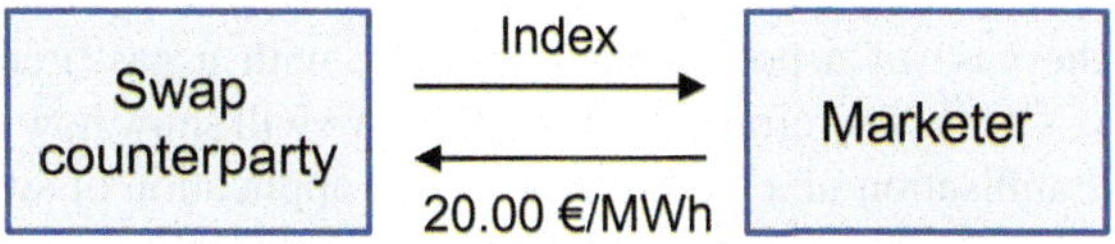

to a counterparty at an index price (right part). This means that the marketer has positioned himself in such a way that he buys gas and sells this gas again directly. He buys at a fixed price and receives an index price—i.e. the volatile market price—for the resale. The figure shows that the marketer receives no net gas.

From the perspective of the marketer, the two trades can be netted into one single transaction. The two physical contract flows balance each other, so that there is no physical flow of energy. What remains is the exchange of a fixed price for an index price. This remainder is essentially what is traded in a swap, which can with one contract replicate what needed two contracts above.

Ultimately, both contracting parties are now simply exchanging payment flows, as Fig. 3.25 shows. There is no more gas flow. The marketer pays a fixed price and receives an index price in return. This is the simplest form of a swap: An exchange of a fixed payment for a variable payment. The swap counterparty in Fig. 3.25 may be a gas producer interested in the opposite: already today agreeing on fixed prices for gas instead of an index. Note again that as the swap is purely financial, the gas producer would sell the gas physically to another party for the index price, but due to the swap is facing no price risk.

Although no physicals are exchanged in a swap, a physical volume must be specified to reference the deal size and to calculate the cash settlement amounts: Both parties must still be able to say to which quantity these payment flows refer. In this example, 20 €/MWh is a reference price that refers to a certain quantity (measured in MWh). It is multiplied with an agreed contract volume. The same applies to the market price paid in the other direction. Furthermore, it must also be determined when the payments are made. Typically, a swap is concluded over a longer period in the future (e.g. several years) and payments are then made regularly (e.g. monthly) within this period.

Figure 3.25 showed a first, simple swap. Two contracting parties (for example a bank and a gas producer) exchange payment flows based on a fixed formula, in this case fixed (20 €/MWh) for floating (index) for an underlying gas quantity. Note however, that the gas producer still sells the gas in the market, receiving the indexed price. Note furthermore that swaps often contain more complicated formulas.

3.3.2.2 Swaps Applied to Generation Assets

Until now, we have discussed fixed-for-floating swaps, i.e. swaps that exchange a fixed payment for a floating, index-priced payment based on one commodity, e.g. natural gas. However, swaps can include more complex payment formulas that include more than one underlying index. This is particularly useful in energy markets where more than just one price determines the profitability of a utility. We present the case of a power plant operator with a gas-fired combined cycle gas turbine (CCGT). Therefore, in this section, we will show how to construct a swap for revenue stabilisation of a CCGT as a typical application of swaps in energy markets.

Section 3.3.1 showed how the profitability of natural gas-fired power plants depends on the (clean) spark spread. Figure 3.23 showed that this clean spark spread can fluctuate over time. The operator of a CCGT may thus want to secure the revenues from the clean spark spread in advance. This can be done with a clean spark spread swap.

Power producers burning fuels face three price risks, which influence their profitability:

1. Power price.
2. Fuel price.
3. CO_2 price.

The three associated price risks can be eliminated by taking the respective positions in the future contracts. However, there may be high transactions costs involved: First, the three commodities would need to be traded in separate forwards. Second, if the respective delivery period is long, this requires numerous forward contracts per commodity. Third, some of these contracts may not be available or illiquid and therefore not be priced properly. In comparison, it can be superior to negotiate one single clean spark spread swap contract.

When constructing a clean spark spread swap, in a first part, the power price risk is eliminated. To do so, the CCGT makes a fixed-for-floating swap. Figure 3.26 shows both the swap (upper part of the figure) and the physical power transaction of the CCGT (right part of the figure). As a result, the generator will sell its power to the physical power market and receive the power index. The swap converts this power index into a fixed payment. The generator's income for the generation of 1 MWh_{el} of power will be the fixed price.

In a second part of the swap, the price risk of natural gas is eliminated using. Figure 3.27 shows the swap and the physical transaction. The CCGT will purchase its natural gas at the physical natural gas market and pay the natural gas index (left part of the figure). In the swap, this gas index will be exchanged for a fixed price (lower part of figure). As a result, the CCGT's cost for receiving natural gas will be the fixed price upon conclusion of the contract. The amount of gas corresponds to 1 MWh/η, where η is the power plant's efficiency. The reason for this is that for generating 1 MWh of electrical power, 1 MWh/η of gas is needed.

The CCGT spread swap now has two parts: One to eliminate power price risk and one to eliminate gas price risk. This is shown in Fig. 3.28.

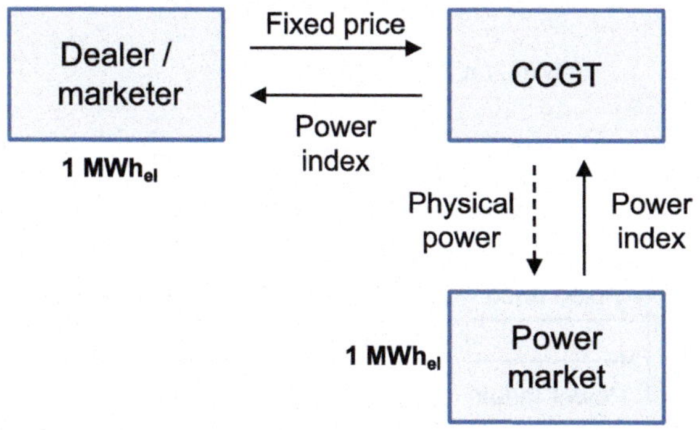

Fig. 3.26 Swap part to secure power price for a CCGT

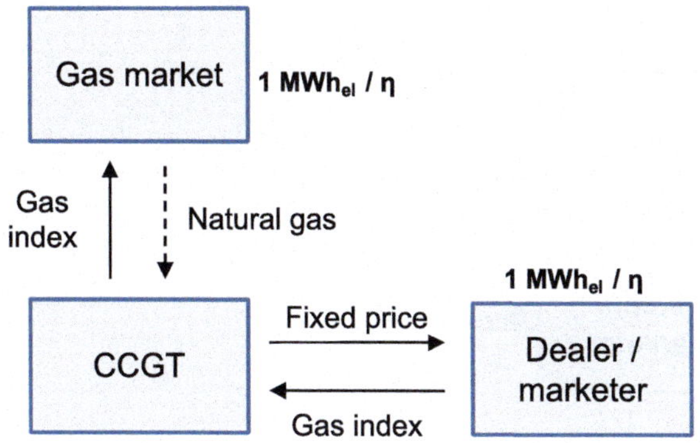

Fig. 3.27 Swap part to secure gas price for a CCGT

In this construct, the CCGT still physically sells its power to the power market and receives the power index (lower part of the figure) while purchasing its natural gas on the natural gas market at the gas index price (upper part of the figure). However, the two financial flow parts are combined in one single swap (left part of the figure). The risk that CO_2 prices will rise in the future and that the power plant can therefore only be operated uneconomically is still there. Because of this, the clean spark spread swap needs a third part to eliminate this risk.

Figure 3.29 shows how CO_2 certificates are purchased and how this shapes the third part of the clean spark spread swap. The CCGT will purchase its CO_2 certificates on the CO_2 market and pay the CO_2 index price (lower part of the figure). In the swap, this (variable) CO_2 index price will be exchanged for a fixed price (left part of the figure). As a result, the CCGT'S cost for receiving CO_2 certificates will be the fixed price agreed on upon conclusion of the contract. The amount of CO_2 corresponds to E_{CO2} multiplied by (1 MWh/η), where E_{CO2} is the carbon emission factor of natural gas, i.e. how many tonnes of CO_2 are emitted when burning one MWh of natural gas. The reason for this is that for generating 1 MWh of electrical power, (1 MWh/η) of gas is needed and by burning this gas, E_{CO2} (1 MWh/η) tonnes of CO_2 are emitted.

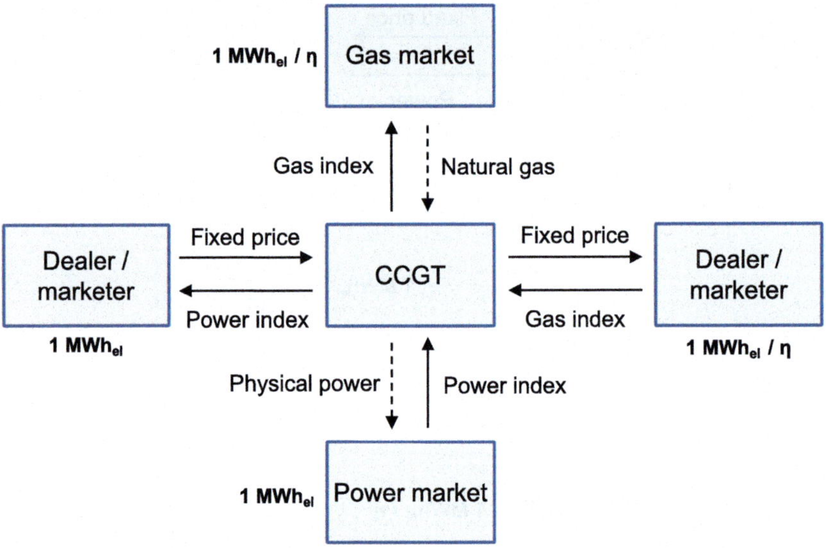

Fig. 3.28 Swap part to secure both power and gas price for a CCGT

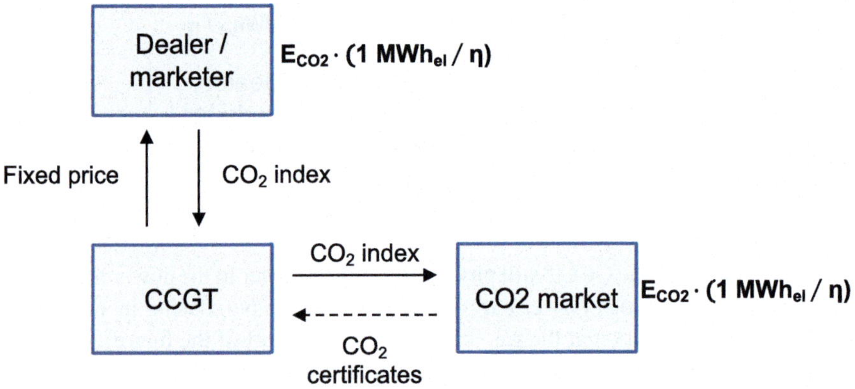

Fig. 3.29 Swap part to secure CO_2 prices for a CCGT

In a final step, the CO_2 part of the swap is added, which leads to the clean spark spread swap.

Figure 3.30 shows the clean spark spread swap and all physical transactions of the CCGT. In this construct, the CCGT still sells its power to the power market and receives the power index (lower part of the figure). It purchases both natural gas on the natural gas market at the gas index price (upper part of the figure) and CO_2 certificates at the CO_2 market (right part of the figure). The three parts are combined in one single swap (left part of the figure).

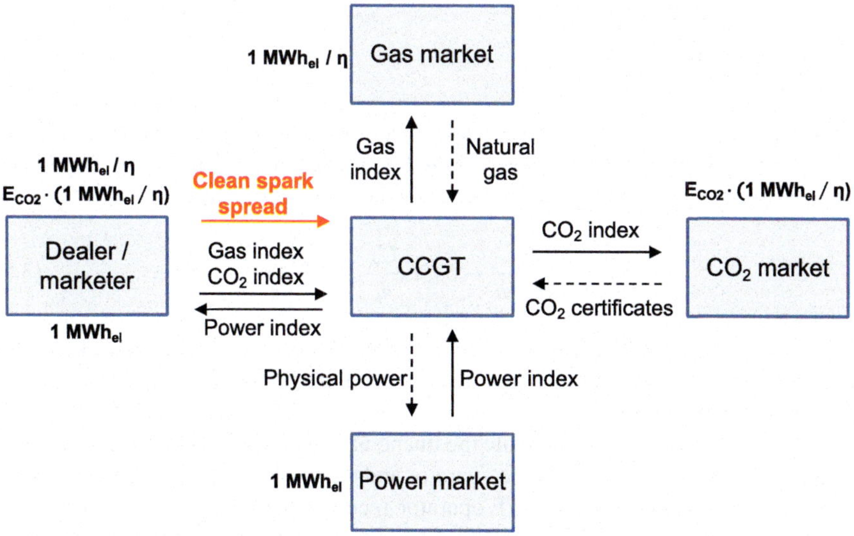

Fig. 3.30 Clean spark spread swap

In this structure of the swap, the CCGT operator exchanges the power index price (which he receives from selling power on the market, thus with a negative sign) and the gas index price (which he needs to purchase gas from the market, positive sign) and the CO_2 index price (which he needs to purchase CO_2 certificates from the market, positive sign). On top of that, he receives the clean spark spread applicable at the time the contract is concluded (positive). This clean spark spread is the difference between the fixed price for power sales and the fixed prices for gas and for CO_2 purchase, adjusted by the CO_2 emissions factor E_{CO2} and the CCGT's efficiency η. Note that the arranged clean spark spread needs to be positive, because otherwise the CCGT would not agree to the swap and refrain from producing electricity.

Furthermore, a physical volume must be specified to calculate the cash settlement amounts. From the perspective of the CCGT, this would be its expected electricity generation during the contract period and the gas and CO_2 emission certificates required to do so. Lastly, a payment schedule needs to be put in place, such as monthly at the first business day of each month for the next ten years.

In the following, we will present an example of a clean spark spread swap, where the operator of a CCGT wants to market its expected monthly generation of 30 GWh while eliminating price risk of power, natural gas and CO_2 certificates for the next year. We will show the details of the swaps and the financial flows associated with it. The clean spark spread swap has the following properties:

- Monthly price settlement from January to December in the upcoming year.
- Fixed clean spark spread of 8 €/MWh.[29]

[29] This is based on the current forward prices for power, gas and CO_2 certificates.

- Power plant efficiency 50%.
- Monthly volume 30 GWh.
- $E_{CO2} = 0.2011$ t/MWh$_{th}$.

As shown in Fig. 3.30, the clean spark spread swap contains a price formula, where the operator receives both gas and CO_2 index prices plus the previously negotiated clean spark spread and pays the power index price in each month:

$$\pi_h = 8\frac{€}{MWh} - S_h + \frac{G_h}{\eta} + \frac{E_{CO2} \cdot C_h}{\eta}. \quad (3.5)$$

Based on this formula, we want to calculate the exemplary financial flows for January and February. In these months, assumed market prices for power, gas and CO_2 are shown in Table 3.5.

These prices allow us to calculate the financial flows between the operator and the swap dealer as well as between the operator and the markets for power, gas and CO_2 certificates. In January, the CCGT operator receives the payments from the clean spark spread swap shown in Table 3.6. Positive values are revenues, negative values are costs.

As a result from the swap, the CCGT operator receives a revenue of €1,446,600 in January. However, the operator also receives money from marketing its power and requires money to pay gas and CO_2 certificates on the wholesale markets. This is shown in Table 3.7.

So in January, the overall benefit of the operator corresponds to €1,446,600 (revenues from the swap) minus €1,206,600 (net payments on the markets), which results in a net benefit of €240,000. Typically, these payments are netted and only the netted amount is transferred between the counterparties of the swap.

In February, market prices have changed, and the operator receives the revenues from the swap that are shown in Table 3.8

Table 3.5 Index price assumptions

	January	February
Power index price P_h	100 €/MWh$_{el}$	103 €/MWh$_{el}$
Gas index price G_h	50 €/MWh$_{th}$	52 €/MWh$_{th}$
CO_2 index price C_h	100 €/t	102 €/t

Table 3.6 Revenues and pay costs for the clean spark spread in January

Clean spark spread	8 €/MWh * 30,000 MWh = + €240,000
Power index price	−100 €/MWh$_{el}$ * 30,000 MWh$_{el}$ = −€3,000,000
Gas index price	50 €/MWh$_{th}$ * 30,000 MWh$_{el}$/50% = + €3,000,000
CO_2 index price	100 €/t * 0.2011 t/MWh$_{th}$ * 30,000 MWh$_{el}$/50% = + €1,206,600
Sum	**+ €1,446,600**

Table 3.7 Revenues and costs on the wholesale markets in January

Power index price	100 €/MWh_{el} * 30,000 MWh_{el} = €3,000,000
Gas index price	−50 €/MWh_{th} * 30,000 MWh_{el}/50% = −€3,000,000
CO_2 index price	−100 €/t * 0.2011 t/MWh_{th} * 30,000 MWh_{el}/50% = −€1,206,600
Sum	**−€1,206,600**

Table 3.8 Revenues and costs for the clean spark spread in February

Clean spark spread	8 €/MWh * 30,000 MWh = + €240,000
Power index price	−103 €/MWh_{el} * 30,000 MWh_{el} = −€3,090,000
Gas index price	52 €/MWh_{th} * 30,000 MWh_{el}/50% = + €3,120,000
CO_2 index price	102 €/t * 0.2011 t/MWh_{th} * 30,000 MWh_{el}/50% = + €1,230,732
Sum	**+ €1,500,732**

Table 3.9 Revenues and costs on the wholesale markets in February

Power index price	103 €/MWh_{el} * 30,000 MWh_{el} = €3,090,000
Gas index price	−52 €/MWh_{th} * 30,000 MWh_{el}/50% = −€3,120,000
CO_2 index price	−102 €/t * 0.2011 t/MWh_{th} * 30,000 MWh_{el}/50% = −€1,230,732
Sum	**−€1,260,732**

Summing up the revenues in February, the CCGT operator receives a slightly higher revenue from the clean spark spread swap, namely €1,500,732. Its revenues and costs on the wholesale markets in February are shown in Table 3.9. Note that they increase by the same amount as the revenues from the swap from January to February.

Hence, the operator's total profit corresponds to €1,500,732 − €1,260,732 = €240,000, the same as in January. This confirms that independent of market prices for power, gas or CO_2, the operator has no more price risk and the operator's profit in each month corresponds to the clean spark spread (in €/MWh) multiplied by the generated amount of power (in MWh). Both costs on the markets for gas and CO_2 and the revenues on the market for power are offset by the corresponding payments as part of the clean spark spread formula.

3.3.3 Options and Generation Assets' Real Optionality

Until now, we have discussed financial instruments with a linear payment structure. This means, that chances and risks are symmetric. Take for instance a utility buying a forward contract. The pay-off structure was presented in Sect. 2.4.3. The company exchanges a variable payment on the spot market for a fixed payment. If the spot price increases, the company benefits to the same amount that it loses when the price goes down—and vice versa. However, such a linear distribution of chances and risks may not always be desired. Instead, one party may want to hedge against rising prices on the purchasing side, but not necessarily benefit fully from falling prices. Conversely, someone on the sell side may want to hedge against prices that are too low but does not need the opportunity immanent in further price rises.

This leads to the concept of options. Generally speaking, an option gives someone the right, but not the obligation, to buy or sell something at a predetermined price at a specific time (or time points) in the future. Since there is no obligation to execute an option (as the term *option* implies), it is only executed when it is beneficial to do so. This gives rise to a nonlinear payment structure. In the following, we will discuss options and optionality, always related to the electricity and gas markets. This includes discussing the different types of options, using options for hedging, and option valuation.

The motivation to discuss options at this place is the analogy to generation assets. Assume you operate a power plant—your pay-off profile is nonlinear! You produce electricity whenever it is favourable and earn a positive revenue. However, when the operation of the power plant would lead to a negative pay-off (i.e. costs for fuel and CO_2 exceed the electricity price) you avoid these losses by simply choosing to turn off your power plant. Hence, your short-term losses are zero while your profits may be substantial. This is a nonlinear pay-off profile like an option. In fact, power plants are said to have "real optionality", a fact we are coming back to in the following sections.

3.3.3.1 Put and Call Options

There are two types of plain vanilla options[30]: call options and put options. The buyer of a call option has the right, but not the obligation, to obtain a certain underlying asset at a predetermined price (strike price) and in a predetermined quantity within a certain period (American options) or at a certain point in time (European options).[31] The seller of the call option is obliged to deliver the underlying asset. As a compensation for this obligation, the seller receives the option premium from the buyer of the option.

Having an option is beneficial because it can be used to create value but does not have to be used if it is not beneficial. If the market conditions do not imply this, the option is not exercised. If the buyer of the option decides not to exercise it, nothing needs to be done. If he wants to exercise it, however, he must actively do it before the expiration date.

[30] "Plain vanilla" refers to the most basic and standard type of financial instrument, without any additional features or complexities.

[31] The terms were coined historically. In modern financial markets, European options can be traded in the Americas and vice versa.

Options as financial products thus have an inherent value because the buyer of an option may execute it if he wants to but does not have to. Therefore, the upside is strictly positive, while the downside is always zero when owning an option (i.e. being long). Consequently, nobody would give the option away for free. Instead, an option has a price, and the seller of the option receives that price for transferring the right described in the option definition to the buyer. The buyer, on the other hand, has to pay that price but afterwards receives a potential payout from the option, which depends on the option definition and the respective market prices.

$$\pi_{\text{LongCall}} = \max\{(S_T - K); 0\} - c. \tag{3.6}$$

Equation (3.6) shows the calculation of the payout π for the buyer of a call option. Here T is the time of maturity (the last date the option can be exercised), S_T is the spot price of the underlying at maturity, K is the strike price (sometimes also referred to as the "exercise price" of the option) and c is the option premium to be paid by the buyer/holder.[32]

The first part of the formula is the spot price minus the strike price. If this is positive, the buyer of the option will exercise the option, because he has the possibility of buying something at a price of K, which can be sold for S_T. As soon as S_T is above K, it is beneficial to exercise the option and the benefit corresponds to the difference between S_T and K. We calculate the maximum between $S_T - K$ and zero, because this difference may also be negative. In that case, the option will not be exercised, and the benefit is zero. Finally, the option premium c must be subtracted, because without receiving a payment for giving away the right to hold this option, nobody will sell it.

The payout profile is shown in Fig. 3.31. On the vertical axis is the payout and on the horizontal axis is S_T. If the underlying is the electricity price, S_T is the electricity price at maturity of the option. In this figure, the price varies from zero to a very large value.[33] If we analyse the profile, we see that it stays at $-c$ for very low prices of the underlying. This is because then the buyer of the option then decides not to exercise it and does not get any money from it. However, the option premium c has already been paid and therefore the total payout is $-c$. As long as $S_T < K$, the option is also referred to being *out-of-the-money*. The payout remains at $-c$ until S_T is equal to K. At this exact point, it does not matter whether the option is exercised or not and it is referred to as being *at-the-money*. If S_T is at least one cent higher than K, the option is *in-the-money* and will be exercised and the buyer receives $S_T - K$, which is positive. For the total payout, c must still be subtracted. So if the option is exercise at a value slightly above K, the total payout will still be negative. Only when the price

[32] In practice, S_T is not necessarily the spot price, because the underlying can be basically any product. In energy markets, the underlying is often a future product, e.g. the year-ahead future. Here, the holder of a call option on the year-ahead future has the right to buy this future for the price K. However, for didactical reasons, we use the spot price in the remainder of the chapter.

[33] Electricity prices can also be negative, but we have yet to see an options with a negative strike price.

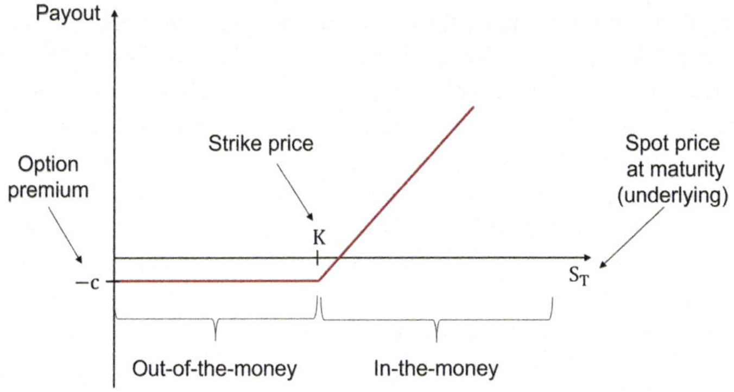

Fig. 3.31 Payout profile for the buyer of a call option

of the underlying is at least $K + c$, the option has a positive payout. Such options can be used in electricity, gas, and even on the CO_2 certificates markets.

Apart from the buyer of the options, there is also the seller who has a payout under certain circumstances. Equation (3.7) describes the payout for the seller of a call option and Fig. 3.32 shows the payout profile. This profile is just the mirrored profile of the payout for the buyer of a call option.

$$\pi_{\text{ShortCall}} = \min\{(K - S_T); 0\} + c. \tag{3.7}$$

Here, if the option is out-of-the-money (S_T is below K), the seller receives the premium c and does not need to do anything else, because the buyer of the option will not exercise the option. However, once the price of the underlying at maturity exceeds the strike price, the seller starts to lose money because the buyer will now exercise the option and the seller has to deliver the underlying at a price K. Since K

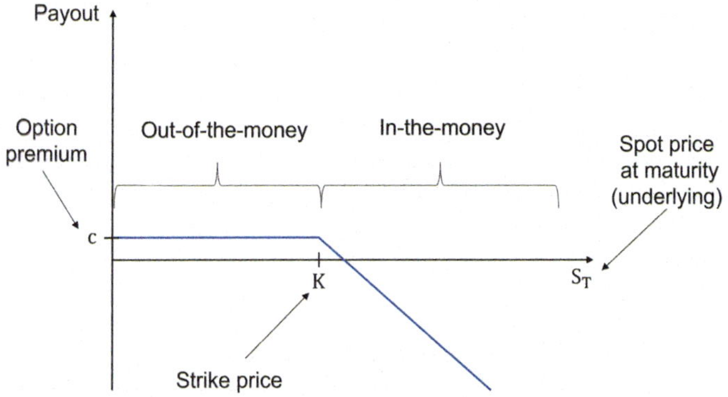

Fig. 3.32 Payout profile for the seller of the call option

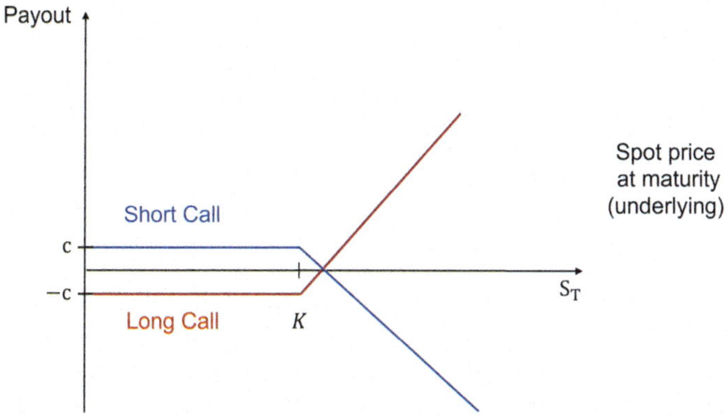

Fig. 3.33 Payout profile of a call option

is now below the market price, this means that there is a loss for the seller and this loss increases linearly with the increase of S_T.

This loss is independent of whether the seller of the option already possesses the underlying or not. If he does, there are opportunity costs, because instead of selling the underlying to the buyer at the strike price K, he could sell it on the wholesale market for a higher price (S_T). If he does not have the underlying, he needs to buy it to be able to hand it over to the buyer of the call option, i.e. he must buy at S_T but has to sell at K. The higher the price of the underlying at maturity, the bigger the loss once the option is exercised.

Figure 3.33 summarises the payout profiles of the buyer and the seller of a call option. It becomes clear that the profit of one counterparty is the loss of the other. The position of the buyer is referred to as *long call*, while the position of the seller is referred to as *short call*.

Call options give the buyer the right to buy the underlying for the strike price. This is particularly useful, if, e.g. the underlying is a fuel that the buyer needs for his generation asset, and he wants to hedge against high prices. However, there may also be situations where not a purchase price shall be hedged but a sales price. This leads to the concept of put options.

Put options work similar to call options: The buyer/holder of a put option has the right, but not the obligation, to sell a certain underlying asset at a predetermined price (strike price) and in a predetermined quantity within a certain period (American options) or at a certain point in time (European options).

The seller of the put option is obliged to buy the underlying asset at the strike price. For this obligation, he receives the option premium from the buyer of the option. From the buyer's perspective, in the worst case he chooses not to exercise this right, and in the best case, if the price of the underlying at maturity is very low, he will exercise it. In this case, it is beneficial to own the put option, because it allows the buyer to sell the underlying at a relatively high price compared to its current market value and hence make a profit. Similar to the call option, the put option's value is positive or zero for the buyer. Therefore, the seller of a put option needs to be compensated with an option premium.

The calculation of the payout of a long put option (the position of the buyer of a put option) is shown in Eq. (3.7), and the resulting payout profile is depicted in Fig. 3.34.

$$\pi_{\text{LongPut}} = \max\{(K - S_T); 0\} - c. \tag{3.8}$$

Figure shows that the lower the price of the underlying, the more profitable is it to own a put option. If the price is zero, the buyer of the put option can and will exercise it, i.e. buy the underlying asset for zero on the market and sell it for K. The option is in-the-money. However, the higher the price at maturity S_T, the less profitable is the option. For instance, assume that the strike price is 20 €/MWh. If S_T is zero, the buyer can purchase the underlying at no cost and sell it for 20 €/MWh by exercising the option. If S_T is now 10 €/MWh, there is still a profit, but it is smaller. Finally, if S_T is 20 €/MWh, the buyer is indifferent between exercising the put and not exercising it. The option is now at-the-money.

At the point where $S_T = K - c$, buyer and seller are both break even. If the value of S_T rises above $K - c$, the buyer still exercises the put option, because it is still in-the-money. However, since he has already paid the option premium c, his total profit will be negative. For prices higher than K, the option is no longer exercised and therefore out-of-the-money. The seller can keep the full option premium c.

The seller's payout profile can be seen in Fig. 3.35. It is derived from Eq. (3.9). The first part of the equation (the minimum) cannot be positive. This means that the seller can either lose money (if the price of the underlying is low) or in the best case get zero. If the value of the first part is negative, it means that the option was exercised and the buyer gets a profit, which translates into a loss for the seller of the option.

$$\pi_{\text{ShortPut}} = \min\{(S_T - K); 0\} + c. \tag{3.9}$$

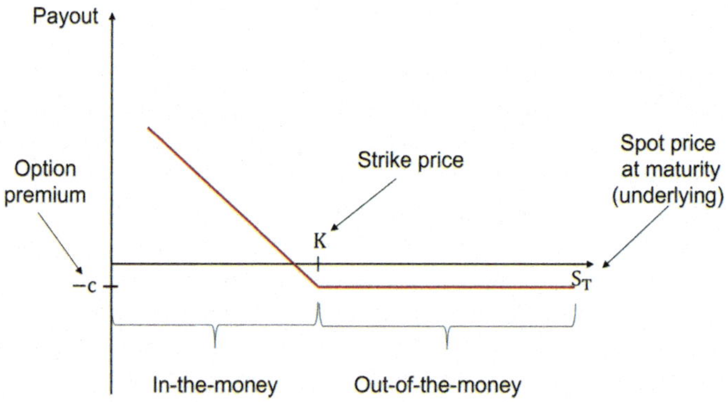

Fig. 3.34 Payout profile for the buyer of a put option

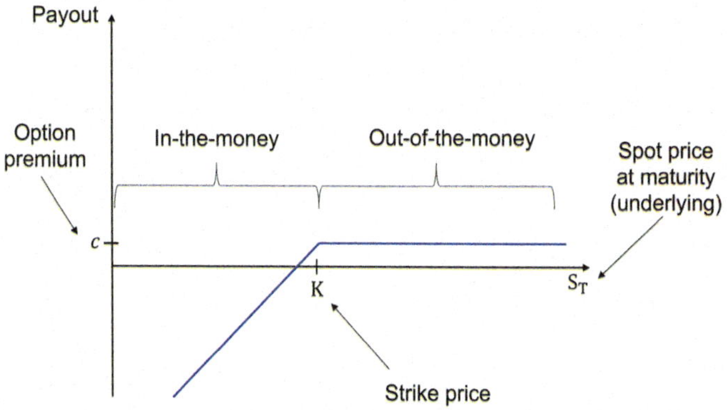

Fig. 3.35 Payout profile for the short put (seller of a put option)

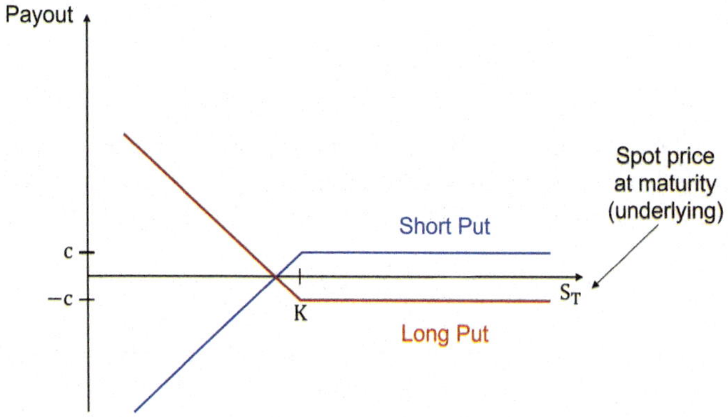

Fig. 3.36 Payout profile of the put option

Figure 3.36 shows the two payout profiles. Like with call options, they are exactly mirrored, i.e. the profit of one counterparty is the loss of the other. The position of the buyer is referred to as *long put*, while the position of the seller is referred to as *short put*.

Table 3.10 summarises some general properties of the two different option types.

3.3.3.2 Spark Spread Options

Simple call and put options are useful tools to structure procurement and sales portfolios and to hedge against unfavourable market developments. More complex options are valuable tools trading the real optionality of power plants. In Sect. 3.3.1, we discussed spark spreads, dark spreads, clean spark spreads, etc. These spreads can also be traded as non-standard electricity options. For example, the buyer of a European spark/dark spread call option written on fuel G (natural gas/hard coal) at a fixed

Table 3.10 Summary of basic properties of a call and put option

	Call option	Put option
Long position: buying of an option	Long call: buying a call option	Long put: buying a put option
Short position selling of an option	Short call selling a call option	Short put: selling a put option
$S_T > K$	In-the-money	Out-of-the-money
$S_T < K$	Out-of-the-money	In-the-money
$S_J = K$	At-the-money	

efficiency η has the right, but not the obligation, to pay at the option's maturity the specified fuel costs and receive the price of electricity.

The payout π at maturity time T for the buyer of such an option is shown in Eq. (3.10),

$$\pi = \max(S_T - G_T/\eta; 0), \tag{3.10}$$

where S_T and G_T are the electricity and fuel prices at time T and η is the efficiency of converting thermal energy to electricity with the fuel. If CO_2 emissions are also considered, the payout π at maturity time T is given by Eq. (3.11),

$$\pi = \max\left(S_T - \left(G_T + E_{CO_2} \cdot C_T\right)/\eta; 0\right), \tag{3.11}$$

where E_{CO_2} is the emission factor, S_T and G_T are the electricity and fuel prices and C_T is the CO_2 emission allowance price at time T.

These equations are in essence a financial replication of a physical power plant's payout profile. If the electricity price is higher than the fuel costs, the power plant should be in operation and revenues are equal to the difference between electricity price and variable costs. If this difference is negative, the power plant should be offline and makes zero profit.

Hence, the physical optionality of a flexible power plant can be replicated by such an option structure. Financially, it makes no difference whether you operate a CCGT or hold a large number of (clean) spark spread options, one for each hour where the CCGT operates.

These options are non-standardised products that are only traded OTC and it would require a lot of effort to fully replicate a power plant's optionality for, say, a year by buying such a large number of (clean) dark/spark spread options. The solution to this is a swing option. Swing options give the buyer the right to exercise an option several times over a certain period (e.g. a year with 8760 hours). Thus, a swing clean spark spread option for one year can be constructed as the financial replication of a CCGT that is in operation for one year. The buyer of this swing option receives the clean spark spread in each hour of the year where it is positive, just like the operator of a CCGT. This has some interesting implications. On the one hand, swing options can be used to value a CCGT. Since the payout of a swing option corresponds to the payout of a CCGT with the same efficiency, its price should also be the same, or at least be similar. On the other hand, if someone wants to buy and operate a CCGT (or a slice of a CCGT), because they want to benefit financially from an expected high clean spark spread in the future, they can instead simply buy a swing option. The financial result is the same or at least similar (because a swing option does not, e.g. account for revision times).

As the swing options are non-standard products, they can contain further restrictions to simulate the physical properties of the power plants even more closely. For instance, a swing option can include a clause that allows the option only to be called a predetermined number of times during a year, e.g. 4000 hours. This makes swing options suitable for evaluating power plants that have, for example, a gas supply contract that stipulates a gas quantity for a certain number of production hours (e.g. between 2000 and 4000 full load hours). Moreover, the swing option may include a clause that replicates revision times, i.e. that the option may not be called for, say, 744 hours in a row during the year.

To sum up, options give you the right to buy or sell something at a predefined price in the future. Plain vanilla options include just one underlying, but there are also customised options that replicate the payout of flexible power plants. This leads to the term real optionality. This and the question when and where to exercise this real optionality in practice which will be explained in the next sections.

3.3.3.3 Real Optionality of Generation Assets

The dispatch of power plants is flexible and spread options can be used to model or evaluate the payout profile of a power plant. This is because the operator of a power plant can decide whether and at what time to produce electricity and thereby has the option to produce or not. This flexibility can be analysed using option theory and power plants' option values are sometimes referred to as **real optionality**.

Valuable flexibility may also exist in the demand profiles of industrial consumers: They decide if they want to consume in a specific hour. Long-term gas supply contracts often allow for shifting procurement between days or even months. Asset portfolios, e.g. a combined heat and power plant (CHP) in conjunction with a peak load heat boiler, heat storage, power-to-heat and/or power-to-gas, give the operator even flexibility on different markets at the same time.

Optionality in the energy sector exists whenever there is the possibility to decide on short notice whether to produce, consume or purchase energy from a contract. This is analogous to what we defined as options in energy: An option gives the buyer the right but not the obligation to buy or to sell something. Power plants can therefore be seen as options and from a financial perspective this technologically complex asset can be converted into a financial product. The analogy between financial and physical asset optimisation is as follows:

- If the clean spark/dark spread is positive in any given hour, the power plant operator will exercise the option and generate electricity (and thus generate a positive profit contribution).
- If the clean spark/dark spread is negative in any given hour, the power plant operator will not exercise the option (and will not generate any profit contribution but will not make losses from short-term operation either).

As a result, the power plant will be operated when it earns a positive profit contribution only, i.e. when the electricity price exceeds short-term variable costs (which depend on fuel and CO_2 emissions costs). Since the electricity price and thus the clean spark/dark spread varies hourly, a flexible power plant corresponds to a series of individual options on the clean spark or clean dark spread, as Sect. 3.3.3.2 showed.

The profit contribution of a power plant in each hour and per MW is given by Eq. (3.12),

$$pc_h = \begin{cases} 0 & \text{, if } S_h < vc_h \\ S_h - vc_h & \text{, if } S_h > vc_h \end{cases}, \tag{3.12}$$

where S_h corresponds to the electricity price and vc_h to the short-term variable generation costs in hour h. These can involve fuel prices like natural gas and CO_2, but also other short-term variable costs.

Figure 3.37 shows graphically when to exercise this option and generate electricity. Here the black line represents the hourly day-ahead electricity prices, and the red line represents the marginal generation costs of the power plant. The plant will generate electricity only when the black line is above the red line, because otherwise generation costs would be higher than revenues for every MWh produced. This is the dispatch decision of a flexible power plant.

Interestingly, this real optionality of the power plant is retained, even if the electricity has already been sold on the forward market. This is because the sale of the forward contract does not require to generate the electricity in the CCGT. While

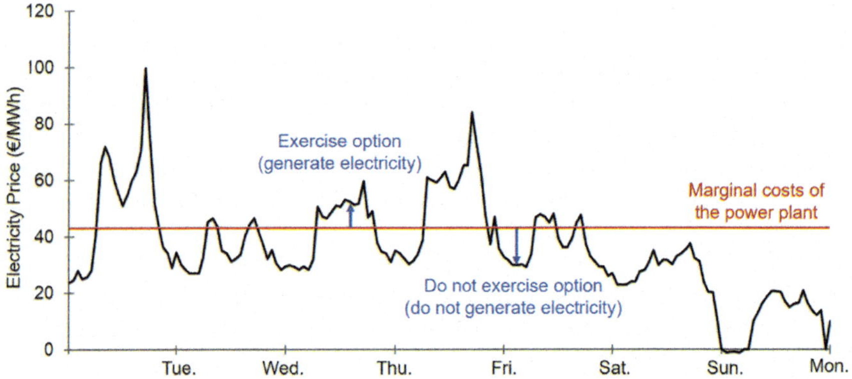

Fig. 3.37 Dispatch decision of a flexible power plant

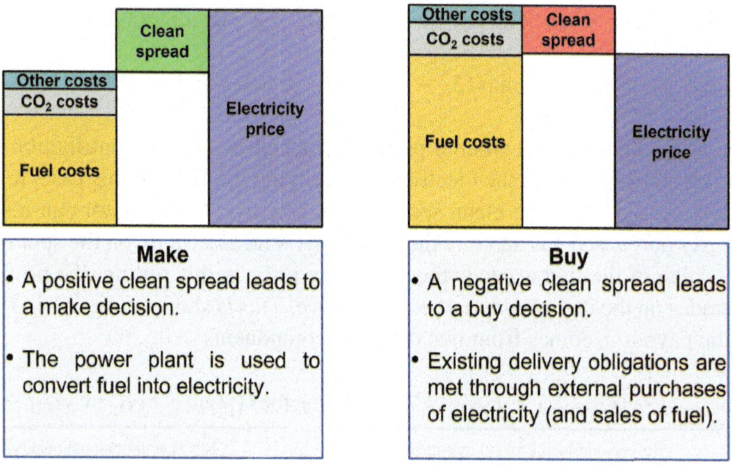

Fig. 3.38 Make-or-buy decision

we *may produce* it in the CCGT, we *may alternatively buy* it on the spot market to fulfil the forward's delivery requirement. The same applies to the fuel bought via a forward contract because it is intended to generate electricity with it. Again, there is no obligation to burn this fuel in the CCGT. If the clean spark spread in any hour is negative, it can simply be sold again.

In other words, as there is no obligation to produce the electricity sold in a forward contract in a generation asset, and there is no obligation to consume the fuel bought in another forward contract (except for some old gas supply contracts where this was explicitly formulated), the power plant operator can still react to short-term price changes. If it becomes cheaper to repurchase the electricity on the spot market due to changes in the prices of the forward before maturity, then this can and should be done. This is referred to as a "make-or-buy" decision.

Figure 3.38 visualises the make-or-buy decision. On the left-hand side, the decision is to produce electricity, because the short-term generation costs (consisting of fuel, CO_2 and other costs) are below the energy price. This decision is based exclusively on the day-ahead market situation and is not influenced by any previous forward agreement. On the right-hand side, the decision is to buy the electricity already sold on the wholesale market, as the short-term generation costs are higher than the electricity price.

Hence, the real optionality of power plants exists regardless of whether forward contracts have been signed or not. However, the specific type of option depends on whether the electricity has already been sold on the forwards market or not. First, we analyse an unhedged plant, i.e. without previous activity on the futures market. An unhedged power plant corresponds to a call option,[34] as we described in the previous section. The production decision for this power plant depends on whether the clean

[34] Or, more precisely, a series of hourly call options.

spread is positive or not. The payout π is in this case given by Eq. (3.13).

$$\pi = \max\big(S_h - (G_h + E_{CO_2} \cdot C_h)/\eta; 0\big). \tag{3.13}$$

The second case arises when a plant has locked-in a profit contribution on the forward market by selling the electricity and buying the fuel. In this case, revenues can be increased when the clean spreads turn negative: the operator can resell the previously purchased gas and buy the previously sold electricity on the spot market. As the value of the fuel exceeds the electricity price in this setting, the profit from these trades on the day-ahead markets is positive. This is shown in Eq. (3.14). In this case, the payout π comes from two different components.

$$\pi = \underbrace{S_{\text{Forward}} - \big[G_{\text{Forward}} + E_{\text{CO}_2} \cdot C_{\text{Forward}}\big]/\eta}_{\text{Revenue through forward transactions}} + \underbrace{\max\big(\big[G_h + E_{\text{CO}_2} \cdot C_h\big]/\eta - S_t; 0\big)}_{\text{Revenue through optionality}} \tag{3.14}$$

The component on the left is the revenue from the forward transactions, which the plant operator has locked-in on the forward markets. This revenue is always positive, otherwise the plant operator would not have made the forward deals. The component on the right is an additional revenue coming from the optionality and is equivalent to a put option: The lower the energy price, the better, because the operator has already sold the energy on the forward market and can buy back the electricity cheaply (and vice versa the fuel).

To sum up, the power plant can exhibit similar properties to a put option when it has been hedged previously, and to a call option when it has not. In both cases, the power plant operator "owns" the option. He can exercise it or not and indirectly pays the option premium with the fixed costs of the power plant.

3.3.4 Short-Term Markets and Flexibility

In the previous sectionss we have mainly dealt with forward markets and how to trade electricity (or gas, oil, coal, etc.) with delivery in the future. The shortest period in the future that we have considered so far has been the following day and thus trading on the day-ahead wholesale market. This is the most liquidly short-term market in Europe and usually sufficient to analyse most trading strategies of energy utilities.

There are, however, also markets that enable energy trading in the very near future. These are referred to as short-term markets and they are typically used to balance portfolios. For instance, in the case of renewable plants' forecast errors it might be necessary to act on these short-term markets, i.e. purchase (or sell) electricity with delivery in a few hours, minutes or even seconds. This is called the demand for flexibility.

As with optionality, we introduce these short-term markets in the section on energy utilities with a generation focus because generation assets provide most of the supply for flexibility. Hence, operators of generation assets often earn additional revenues from trading on short-term markets. This section first explains the markets for flexibility and how they function. The following section shows how to optimise generation assets to generate additional revenues on these markets.

Until now, we have referred to the day-ahead market as a spot market. The reason for this is that this market has a high liquidity and is used to balance portfolios reported to the TSO on the day before delivery. However, there are additional markets with even shorter time horizons: the intraday market and the balancing market. The latter usually consists of three sub-markets. In most countries, the balancing market is divided into:

- Frequency containment reserves (primary reserve).
- Frequency restoration reserves with automatic activation (secondary reserve).
- Frequency restoration reserves with manual activation (minutes reserve).

The actual segmentation and specific regulation of balancing markets depends on each country's specific need for short-term flexibility. In Sect. 3.3.4.2 we will give examples of such country specific regulation. Nevertheless, the abovementioned three markets are a good generalisation of balancing markets and most countries' regulation is based on them.

3.3.4.1 Intraday Markets

Intraday trading is continuous and takes place 24 hours a day and 7 days a week. Trading for the following day typically starts on the afternoon, and trades can happen until 5 min before physical delivery.[35] There are usually hourly, half-hourly and quarter-hourly products. Additionally, since trading is performed continuously, there is no uniform market price on the intraday market. Transactions are settled immediately as soon as the seller and the buyer have agreed on a price.

The intraday market opens after grid closure. Before grid closure, balancing group operators had to report their expected supply and demand in their balancing group for the next day to the best of their knowledge. Therefore, intraday markets should be used to trade changes in the forecasts becoming known after grid closure. Consequently, the intraday market is not relevant for a pure financial balancing group. This is because there are no deviations in such a balancing group since it is not going to consume or produce electricity. The intraday market is therefore only relevant for market participants with physical assets such as power plants, consumers, etc.

[35] At least, this is possible via the EPEX spot platform as of May 2023. Conditions such as lead time may vary from exchange to exchange and from TSO to TSO.

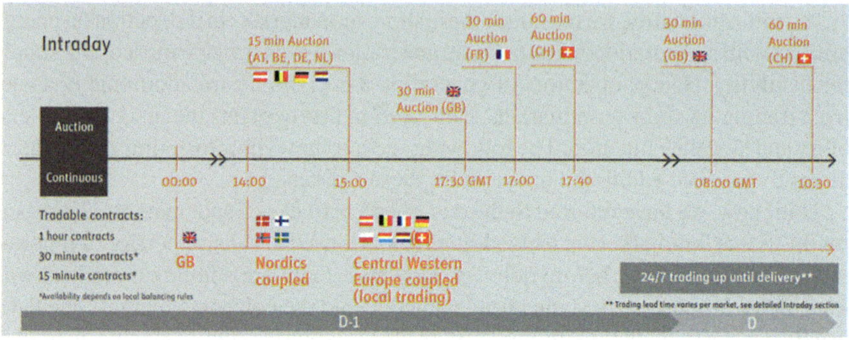

Fig. 3.39 Timetable for intraday trading in European Countries[36]

Figure 3.39 shows in more detail the timetable for intraday trading in Europe:

1. In most countries, at 15:00, there is an auction for the 96 quarter-hours of the following day.
2. Depending on the country, there may be further intraday auctions after that.
3. Continuous trading for most contracts starts at 15:00.
4. Trading time end shortly before physical delivery, in most markets only a few minutes.

It is remarkable how much work the transmission system operators must do to keep the system stable and without blackouts, considering how close to delivery trading takes place and how complex the physical system is.

Figures 3.40 and 3.41 show a sample development of intraday prices for delivery on one day. Red is the highest price that was paid for a certain product, green the lowest price and in grey is the weighted average price. Also, the ID1 (the weighted average price for all trades executed within the last trading hour of a contract) is included. The quarter-hourly intraday prices (Fig. 3.41) are more volatile for the same product than hourly intraday prices (Fig. 3.40). Also, price range for the same product can be very high. For instance, the price for the hour from 1:00 to 2:00 lied between 0 and about 70 €/MWh depending on when the trade was performed. For quarter-hourly products this price range is a lot higher. Hence the time of trading is important for an optimal trading strategy. Because of that, additional profit possibilities arise for the owner of a flexible power plant. The operator of the power plant may sell, rebuy or resell the exact same product (e.g. delivery from 1:00 to 1:15) several times on the intraday market. Since the flexible power plant can—but does not have to—physically deliver the energy, the make-or-buy decision process previously explained may be executed several times. If the price for a particular product evolves as shown in Fig. 3.42, this can lead to high profits. By selling the product when the price is high and buying it back when it is low, revenues increase with each deal, even in the case where no energy will be delivered in the end. Note however that technical properties of the

[36] EPEX Spot (2022).

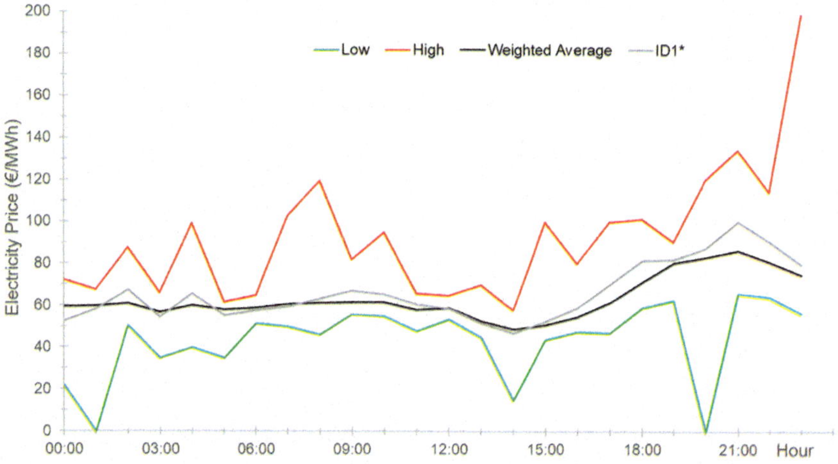

Fig. 3.40 Example of hourly intraday prices

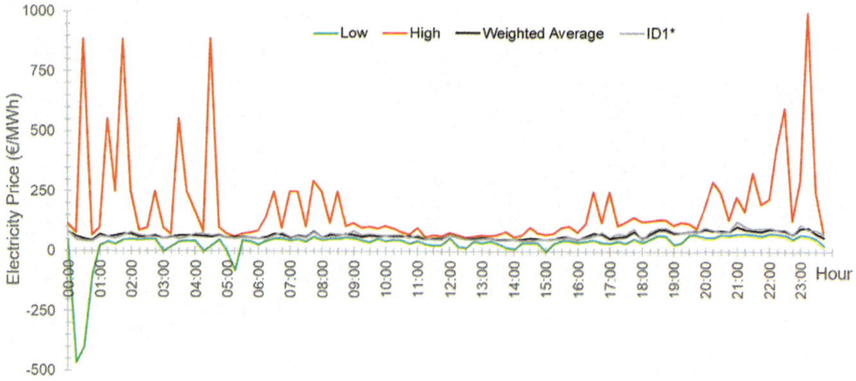

Fig. 3.41 Example of quarter-hourly intraday prices

power plant (in particular, ramp rates and minimum load requirements) may provide binding constraints for such a strategy.

Intraday markets are becoming more and more important. In Fig. 3.43 we can see a comparison between the intraday markets and the day-ahead market in terms of trading volume between 2015 and 2021 in the markets traded at EPEX SPOT. On the day-ahead market, the total trading volume remained almost the same in all countries. While intraday market liquidity has more than doubled since 2015, it still lags behind day-ahead markets (note the differences in scale between left and right graph).

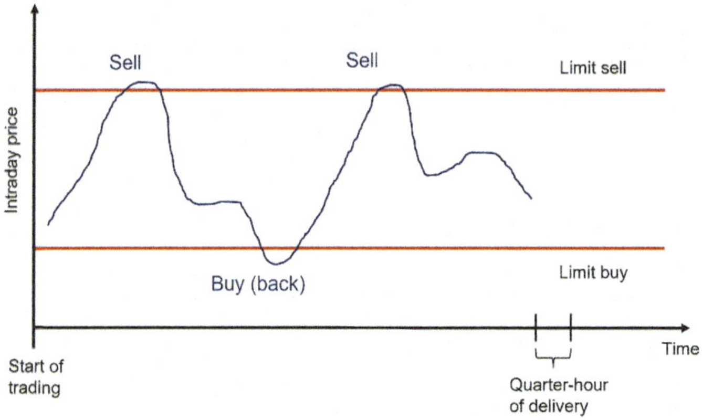

Fig. 3.42 Example of the trading strategy of a flexible power plant on the intraday market

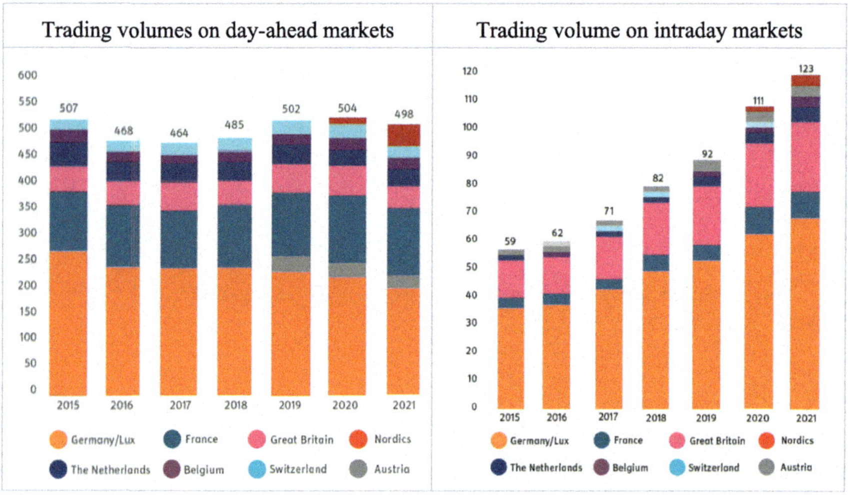

Fig. 3.43 Day-ahead market and intraday market liquidity in EPEX SPOT markets[37]

3.3.4.2 Balancing Markets

We have already mentioned the importance of a stable frequency for the electric system. For a stable frequency, generation and consumption need to be equal in real time. In Europe, the nominal frequency's target value is 50 Hz,[38] but any deviation between generation and consumption, in every second or even millisecond, will let the frequency deviate from its nominal value. These frequency deviations must

[37] EPEX Spot (2022).

[38] In other countries, such as, e.g. the USA, it is 60 Hz.

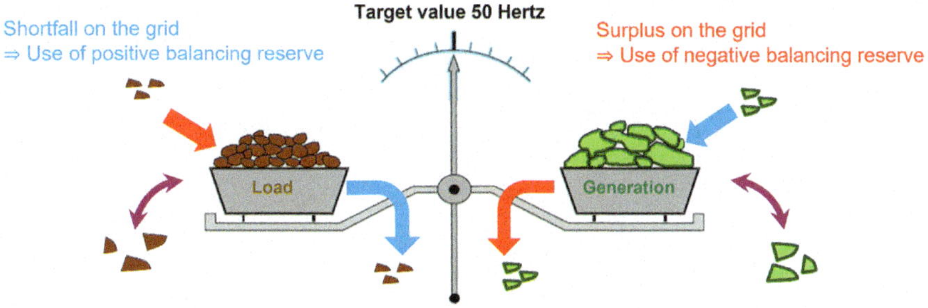

Fig. 3.44 Effect of generation and load on system frequency

be corrected to maintain the quality of supply, as well as system reliability. Small deviations are normal and allowed, but high deviations must be corrected.

To bring back the frequency to its target value, positive balancing power is used if the frequency is too low and negative balancing power if the frequency is too high. These adjustments are organised centrally by the transmission system operator (TSO), based on previously contracted capacities of different qualities in the form of ancillary services.

In Fig. 3.44 we can see a graphical representation of the relationship between load, generation, and frequency. If the load exceeds the supply, the frequency drops (the scale arrow moves to the left) and must be compensated. This can be done in two ways: Either consumption must be decreased, or generation needs to be increased. If, on the other hand, the frequency is too high because there is more generation than consumption, we have again two possibilities: increase consumption or reduce generation.

As a result, balancing services can be provided both from the demand and from the supply side, and both can earn revenues from it. In the following, we will focus on the generation side, but the same principles apply to the demand.

In addition to the positive and negative distinction of reserves, contracted capacities are also differentiated in terms of quality (especially the duration until availability). Typically, in most markets there are a total of five different products on the market. We can see the requirements for the different products in Fig. 3.45.[39]

The red curve represents the frequency containment reserve ("FCR"). FCR is symmetrical, which means that it must be provided both positively and negatively at the same time. Consequently, there is only one product for this type of reserve. This reserve must be available very soon. It is activated immediately after a frequency disturbance and must be fully provided within 30 s after activation. The blue curve represents the automatic frequency restoration reserve ("aFRR"). It can be positive or negative, which means that there are two different aFRR products. It must be activated 30 s after the TSO's signal and be fully available after 5 min. Finally, in

[39] This is the current regulation that applies to most countries in the European Union. The specifics of the different products can vary from country to country, but the principle applies to most countries with balancing markets.

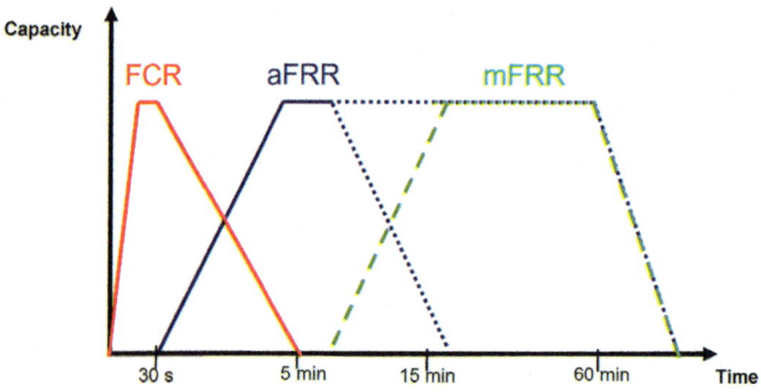

Fig. 3.45 Different balancing reserve products

green, we can see the manual frequency restoration reserves ("mFRR"). This can also be positive or negative (two products). Its activation must start about 7.5 min after the TSO's signal (depending on the schedule) and must be fully available after 15 min.

Originally, these products (FCR, positive and negative aFRR, positive and negative mFRR) were managed independently by each TSO. However, this leads to inefficiencies in procurement and activation. For instance, it may happen that in two neighbouring control areas, aFRR is activated simultaneously in the opposite direction, because there is a power surplus in one region and a deficit in the other. Although in such a situation it would be economically efficient to balance only the net deficit of the two regions, this is not the case if the two TSOs act independently. Therefore, in Europe, the European Commission established a guideline on electricity balancing in November 2017: Commission Regulation (EU) 2017/2195.[40] The goals of this regulation are a common market for procurement and exchange of balancing services to foster effective competition, non-discrimination, transparency, new entrants and increased liquidity. A European implementation project exists for every balancing service, to define technical, operational and market rules. The current status of these projects will be presented later in this section.

The necessity of balancing services derives from the frequency variations, which are provoked by several factors:

- Renewable energy feed-in may be higher or lower than the forecast used for balancing the system at grid closure on the day before delivery.
- The load may be higher or lower than forecasted at grid closure.
- Power plant outages may reduce the supply (In this case only positive balancing power is required.).
- Averaging of short-term supply and demand. For instance, if the system is organised in quarter-hourly intervals, in the mornings when the load is increasing, during the first minutes of the interval, the demand is below the average, and during the last minutes, the demand is above the average. As a result, negative

[40] European Commission (2017).

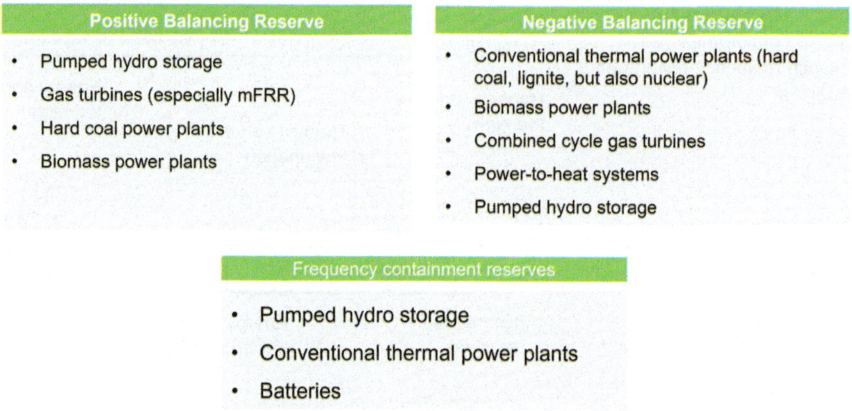

Fig. 3.46 Typical technologies providing the different reserves

balancing power is required during the first and positive balancing power during the last minutes to manage these differences.

- International electricity exchange might also drive frequency deviations.

The amount of required balancing reserve is determined with probabilistic models like stochastic convolution. In the following, this process is assumed to be exogenous.

Not all power plants (or consumers) are equally suited to provide the different reserves. Technical characteristics of the plant must be considered, but also the exact bid structure must be estimated. This requires complex models, which are beyond the scope of this book. In Fig. 3.46 we can see a list of the technologies that typically provide the different reserves services.

For the allocation and remuneration of the automatic and manual frequency restoration reserves, the prospective suppliers must be prequalified, i.e. they have to be tested to guarantee that they comply with the technical requirements, e.g. that their response to the command signal is fast enough. Afterwards, the prequalified suppliers may bid for the service and become part of the next day's manual and automatic frequency restoration reserves.

First, the bidding is done in a so-called capacity market. This market serves as a backup for the TSOs, as the capacities contracted here must be fully available during the market time unit (MTU). Market time unit means the period for which the market price is established. On balancing power markets this may last from a few minutes up to several hours. It can be compared with the delivery period for future products (see Sect 2.1.2). The submitted bids include the offered capacity (in MW) and the capacity price (in €/MW). Once all bids are submitted, the bids are sorted according to the capacity price and bids are accepted starting with the lowest bids until all required capacity is procured. The offered capacity is remunerated with the offered capacity price, i.e. it is a "pay-as-bid" market.[41]

[41] It is an ongoing debate whether this is optimal or whether a uniform price procedure would not be a better solution. Thus, in some countries, "uniform pricing" may be used instead of "pay-as-bid".

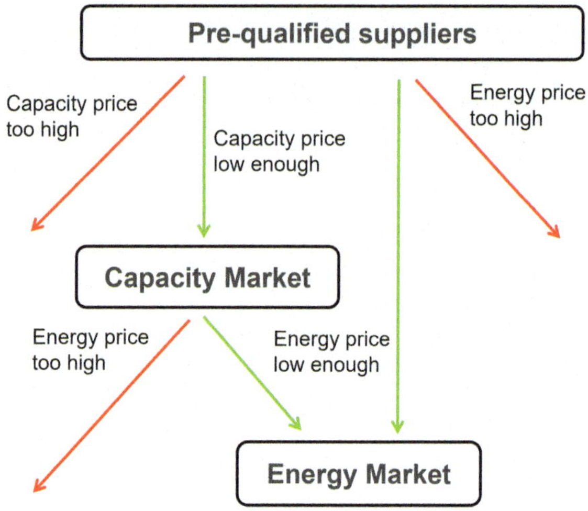

Fig. 3.47 Allocation process of automatic and manual frequency restoration reserves

As a result, some of the prequalified suppliers will enter the capacity market if their price is low enough, and some will not, if their bid is too high, or if they did not submit any bid. From this, we get the total guaranteed capacity for the TSO to balance the system in real time. The capacity market signals that a capacity provider with a flexible system would be prepared to supply energy at short notice if the TSO so wishes. However, the remuneration for the actual provision of energy has not yet been determined.

For this, there is a second market behind the capacity market, the so-called energy market. In this market, the winners of the capacity market auctions must submit bids. Additionally, bids of all prequalified suppliers are also allowed. Even if a potential supplier did not make it into the capacity market, they are still allowed to submit energy bids on the energy market. However, they are not obliged to do so and they will not receive any payment for the capacity like the winners of the capacity market auction.

The bids on the energy market include an offered capacity (in MW) and an energy price (in €/MWh). The TSO then sorts all bids according to the energy price in ascending order. This is the merit order for the energy market for balancing power for automatic and manual frequency restoration reserves. Whenever balancing energy is required, bids are called according to this merit order starting with the cheapest. The remuneration for activated energy is according to the bided energy price. This means that this is also a "pay-as-bid" market (balancing markets tend to be "pay-as-bid" markets and day-ahead markets tend to be "uniform price").[42] The allocation process of automatic and manual frequency restoration reserves is shown in Fig. 3.47.

[42] Again, there is an ongoing debate whether this should be uniform pricing instead.

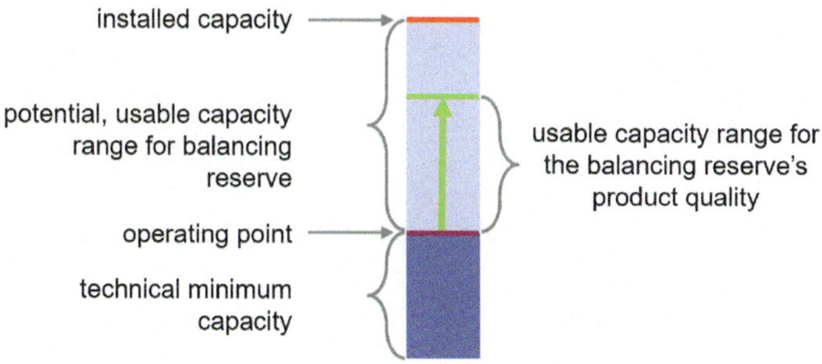

Fig. 3.48 Maximum supply quantity of positive reserve

The provision of positive balancing power typically means that the generator ramps up to provide the service, increasing generation from the levels already committed in the day-ahead market. The maximum supply quantity a power plant can provide depends on the capacity which is kept free between the technical minimum capacity and nominal capacity. For instance, if an 800 MW coal-fired power plant is producing 700 MW, this plant could sell at most 100 MW on the positive reserve market.

An additional constraint for the maximum capacity a plant can provide in the reserves market comes from the ramp rate of the plant. This is the technically possible power output increase or decrease in the relevant time range of a product. The ramp rate is usually given in megawatt per minute increase or decrease. If the coal-fired power plant mentioned above can only ramp up 80 MW during the activation period, then this is the maximum value of the reserve that the power plant can provide.

In summary, the maximum amount of reserve that a conventional power plant can provide is the minimum between the capacity kept free between the technical minimum and the nominal capacity, and the technically possible increase in the relevant time range. This is shown in Eq. (3.15) and Fig. 3.48.

$$CAP^{BR+} = \min\{CAP^{EL} - COP^{EL}; \Delta CAP^{EL}\}, \tag{3.15}$$

where

$\boldsymbol{CAP^{BR+}}$: Maximum amount of positive capacity that the plant can provide in a certain hour.
$\boldsymbol{CAP^{EL}}$: Installed capacity.
$\boldsymbol{COP^{EL}}$: Current operating point of the power plant.
$\boldsymbol{\Delta CAP^{EL}}$: Maximum power increase of the power plant in the activation time.

The maximum negative reserve supply quantity is the minimum of the capacity kept free between the technical nominal capacity and the technical minimum capacity, and the technically possible capacity decrease in the relevant time range.

In the case of the coal-fired power plant generating 700 MW, if the minimal technical capacity is 300 MW, we could decrease generation by 400 MW. But again,

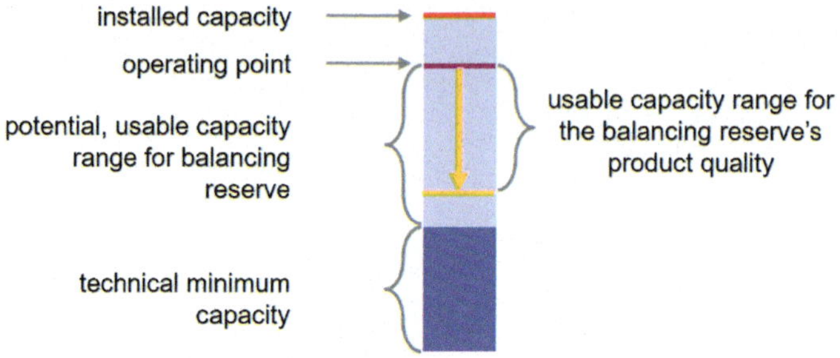

Fig. 3.49 Maximum supply quantity of negative reserve

we need to consider that the ramping constraint may impose an additional limitation depending on the ramp-down rate and the time duration of the respective product.

This is shown in Eq. (3.16) and Fig. 3.49.

$$\boldsymbol{CAP^{BR-}} = \min\{\boldsymbol{COP^{EL}} - \boldsymbol{CAP^{EL,Min}};\ \boldsymbol{\Delta CAP^{EL}}\} \tag{3.16}$$

where:

$\boldsymbol{CAP^{BR-}}$:	Max. amount of negative capacity that the plant can provide in a certain hour.
$\boldsymbol{CAP^{EL,Min}}$:	Technical minimum capacity.
$\boldsymbol{COP^{EL}}$:	Current operating point of the power plant.
$\boldsymbol{\Delta CAP^{EL}}$:	Maximum power decrease of the power plant in the activation time.

If we are in charge of the operation of a power plant or involved in market design, then we need to know about the different approaches to determine the level of balancing reserve price. If we are considering investing, buying or selling and actively participating in positive and negative reserve power markets, we must understand how prices are formed and what drives these prices. This depends on the specific situation, and there are four possibilities, two for positive balancing reserves and two for negative balancing reserves.

The remainder of this section shows how the costs for providing balancing power capacities can be determined. It is based on the analysis in Müsgens et al. (2014). These analyses are focused on spinning reserves, and hence do not include, e.g. wind turbines or emergency standby power.

Cost of provision of positive balancing reserve

- **The asset is in-the-money (**S_h**>vc)**. If the price of electricity is above the power plant's variable costs, by providing positive balancing reserves we face opportunity costs because we need to block capacity for balancing, and this reserved capacity does not receive revenue on the wholesale market.

  Following the previous example of our 800 MW coal-fired power plant, we can produce and sell 800 MW on the wholesale market, because the price is attractive, and we can earn revenue. However, if we already sold 800 MW and that is our maximum capacity, we cannot increase our output any further, and thus, we cannot provide positive balancing power. This means that we have to decide if we sell each megawatt on the wholesale market, or we block it for providing balancing power.

Additionally, we face increased specific generation costs, due to the loss of efficiency during the partial load operation. If we reserve capacities for balancing power, the plant cannot produce at 100 %, and this leads to a small, but non-negligible loss in efficiency.

- **The asset is out-of-the-money (S_h<vc).** If the variable cost is higher than the electricity price, producing electricity will generate revenue, but this revenue will be lower than the variable production costs, and we have losses. Nevertheless, some plants need to be generating electricity to be able to sell positive balancing power, because otherwise, it is technically impossible to start them up within the required activation response times for balancing power.[43] Some plants like open cycle gas turbines (OCGT), hydro plants, and batteries can be started within a few minutes. These plants can avoid this problem when providing certain balancing power products.

  If the plant needs to be running with at least minimal load, the plant incurs losses on the wholesale market. If the minimal load of our coal-fired power plant is 300 MW, this power needs to be sold on the wholesale market knowing that the revenues on the day-ahead market do not cover the variable costs of producing this energy. Therefore, the revenue for providing balancing power must exceed the loss incurred from the day-ahead operation.

Let us look at the cost structure incurred for providing positive balancing reserves for different technologies. In Fig. 3.50, on the left-hand side, we show technical data of a typical hard coal plant, a typical combined cycle gas turbine, and a typical biomass power plant. This includes typical nominal capacities and variable generation costs of each plant. Also, we assume that the biomass plant receives a market and management premium as a subsidy.

On the right-hand side of Fig. 3.50, we see graphically how much it costs to provide balancing reserve in € per MW per hour (vertical axis) as a function of the spot price at the wholesale market (horizontal axis). Here we can see the implications of what we explained: If prices are lower than variable costs but plants must be online, they incur costs, and the lower the spot price the higher the costs. Additionally, when the electricity price is higher than variable costs, they incur opportunity costs. These costs increase when the spot price rises.

[43] See the case studies at the end of this section for details on the required response times. Depending on the balancing product, these can be just a few seconds up to several minutes.

Assumptions – Power plant technologies

	Hard-Coal PP	CCGT	Biomass
Variable generation cost in €/MWh	40	60	80
Market and management premium in €/MWh			50
Minimum capacity in %	30%	40%	50%
Nominal capacity in MW	800	400	100
Balancing Reserve in MW	100	100	50

Positive balancing reserve

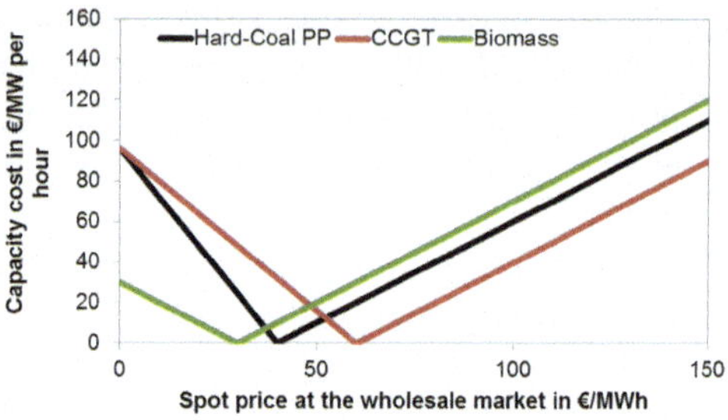

Fig. 3.50 Capacity cost of providing positive balancing power for different technologies[44]

The influencing factors that drive these V-shaped cost curves are the electricity price, variable generation costs, minimum capacity, and balancing reserve slice (depending on capacity gradient and product). Other factors that were not considered in this model are efficiency loss during part-load operation and start-up costs.

As the cost of providing the balancing reserve depends on the electricity price, efficient suppliers vary with the electricity price. This also means that they change over time with changes in fuel and CO_2 certificate prices. For instance, if the spot price is slightly below 50 €/MWh, we see in the graph that hard coal has the lowest cost to provide balancing power.

Provision of negative balancing reserve

- **The asset is in-the-money (*S_h*>vc)**. If the power price is higher than variable cost, the power plant should operate at maximum load to earn as much money as possible on the day-ahead market. Providing negative balancing power means that the power plant must be able to ramp down. If we are producing at maximum load, no change of the operating point is necessary to provide negative balancing power. We just continue to do what we had planned on the day-ahead market to make profit. Therefore, providing negative balancing reserves can be done at no cost.
- **The asset is out-of-the-money (*S_h*<vc)**. This situation is similar to when the plant is out-of-the-money and providing positive balancing reserve. Just as in this case, the plant must be running even if it makes losses. The difference to providing positive balancing reserve is that the plant must now generate more than the minimum load to be able to ramp down. Therefore, so the losses are even higher.

In the following we will analyse the cost structure incurred for providing negative balancing reserves. The power plants and their technical parameters are the same as in the case of positive reserves (see Fig. 3.50). The resulting capacity costs are shown on the right-hand side of Fig. 3.51.

[44] Müsgens et al., (2014).

Assumptions – Power plant technologies			
	Hard-Coal PP	CCGT	Biomass
Variable generation cost in €/MWh	40	60	80
Market and management premium in €/MWh			50
Minimum capacity in %	30%	40%	50%
Nominal capacity in MW	800	400	100
Balancing Reserve in MW	100	100	50
Negative balancing reserve			

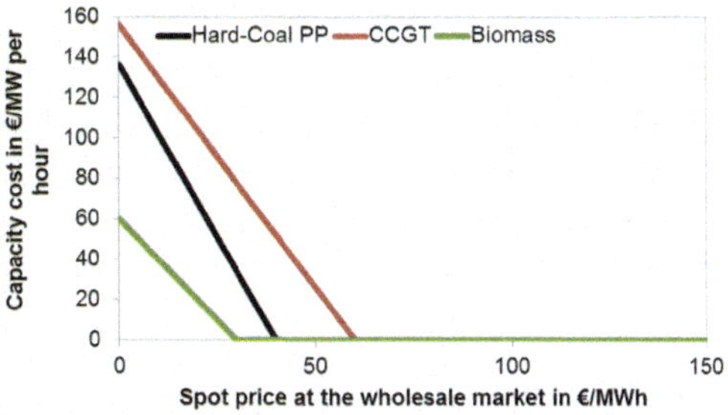

Fig. 3.51 Capacity cost of providing negative balancing power for different technologies[45]

As we explained, if these technologies are in-the-money, then they should operate at maximum capacity and there is no opportunity cost for providing negative balancing power. If the spot price is for instance 150 €/MWh, the cost of providing negative balancing power is zero for all three technologies, because they do not need to make any changes on the day-ahead market.

However, if the wholesale price for electricity is below the production costs, the power plants must continue to produce on the day-ahead market, even though this leads to losses. As can be seen in Fig. 3.51, this quickly becomes expensive as the curves are steeper than with positive balancing power.

The influencing factors are the same as with positive reserve. When plants are out-of-the-money, different plants have different costs for the provision of balancing power at each electricity price. Therefore, the "efficient" suppliers vary with the electricity price and over time with changes in fuel costs and CO_2 certificate prices.

This characteristic of negative balancing reserves is increasingly becoming an issue since there are more and more renewables in the systems. They replace conventional power plants but have limited ability to provide balancing power.

Apart from the abovementioned general assessment on the cost structure of spinning reserves for providing balancing power, there are further aspects influencing this cost structure:

- Duration of the market time units: Due to electricity price variations, a power plant may be both in-the-money and out-of-the-money during different hours of the same market time units. The shorter the market time units is, the less of a problem this is.
- Dynamic aspects: Start-up and shutdown costs, minimum downtime and minimum operating times need to be taken into account as well.
- For combined heat and power (CHP) plants: It is necessary to consider the heat side of the CHP plant as well. This means that when changing the operation point due to the provision of balancing power, the loss or surplus of heat and its value also needs to be considered.

[45] Müsgens et al. (2014).

- For consumption units: The costs in the production process due to the fluctuating energy supply must be considered.
- For power plants and consumers with connection to high, medium, and low voltage networks: Avoided grid usage charges may need to be considered.[46]
- Other aspects: Alternative revenue options, such as intraday markets, need to be considered. If there are several assets that can be combined, portfolio effects may be lost, which causes costs.

Price formation mechanism in balancing reserve markets

We have seen how different technologies can provide balancing services and how their costs depend on the wholesale price. We also know the demand for balancing power (calculated by the TSOs), but we still do not know what the balancing reserve market price will be in different situations. In the following, we will analyse fundamentally, how the price of balancing power can be estimated. To do that, we have to make the following assumptions:

- There is perfect competition (no consideration of market power).
- The technical properties of conventional power plants need to be simplified:
  - No consideration of start-up costs.
  - No consideration of efficiency loss during part-load operation.
  - Uniform assumptions on technical minimum load and load gradients for each technology.
- There is no consideration of storage and pumped hydro storage power plants, consumption and fluctuating renewable energy units as potential bidders.
- Positive and negative balancing reserves are analysed independently of each other.
- There is no consideration of a call of balancing energy (and the respective revenues).
- There is no consideration of marketing and bidding strategies.

These simplifications should not lead to fundamental changes in the derived results. In the following, we will perform an example calculation to get an idea of how the balancing reserve market works, and how prices are established.

The example calculation is shown in Fig. 3.52. In this case, we assume that there are five generation technologies (nuclear, lignite, hard coal, CCGT and gas turbines). The resulting merit order on the day-ahead market is shown on the top left of the top left chart of the figure. We also assume that there is a system load of 45 GW. The intersection of supply and demand in the wholesale market gives the wholesale market price of 44.46 €/MWh. We can determine the inframarginal power plants: nuclear, lignite, and approximately half of the hard coal. The other half of the hard coal, the CCGTs and the open cycle gas turbines are extramarginal and are hence not generating. This is the situation without taking account of balancing power.

[46] This depends on the regulation in the country where the unit is located.

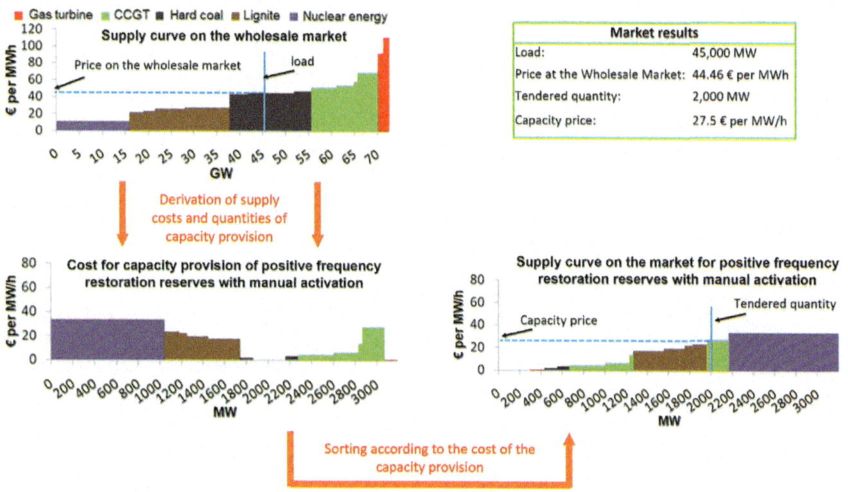

Fig. 3.52 Supply curve derivation for positive mFRR for a total demand of 45 GW[47]

Now we will derive the costs of positive balancing power (mFRR). To do that, we assume that only some of the power plants of each technology can actually provide balancing power. We also assume that 2000 MW of balancing power are required. Since we already know the wholesale market price, we can calculate each power plant's cost for providing balancing power using the methodology outlined in Figs. 3.50 and 3.51. This leads to a V-shaped curve in the bottom left chart of Fig. 3.52. Here we can see that the costs of providing balancing power for hard coal are very low, because hard coal plants are setting the price, i.e. are at-the-money. Some of the hard coal plants are inframarginal and some of them are extramarginal. In all cases, however, their generation cost deviate only slightly from the wholesale market price. The inframarginal ones make a very small profit contribution and the extra marginal ones would not be losing much if they decided to generate electricity and sell it on the day-ahead market. Therefore, their costs for providing balancing power are low.

Lignite power plants, which are inframarginal, would need to reduce their output to free capacity and provide positive balancing power. This, however, would mean losing revenue on the day-ahead market. The same applies to nuclear power plants, with the difference that they would lose even more revenue because their variable production costs are very low. For CCGTs, which are extramarginal, it would be costly to enter the market, which translates into high capacity costs for providing balancing power. The capacity costs for gas turbines are zero, as their response time is so short that they can provide positive balancing power without being in operation. When they are extramarginal, they simply stand still and can provide balancing power without incurring costs.

---

[47] Müsgens et al. (2014).

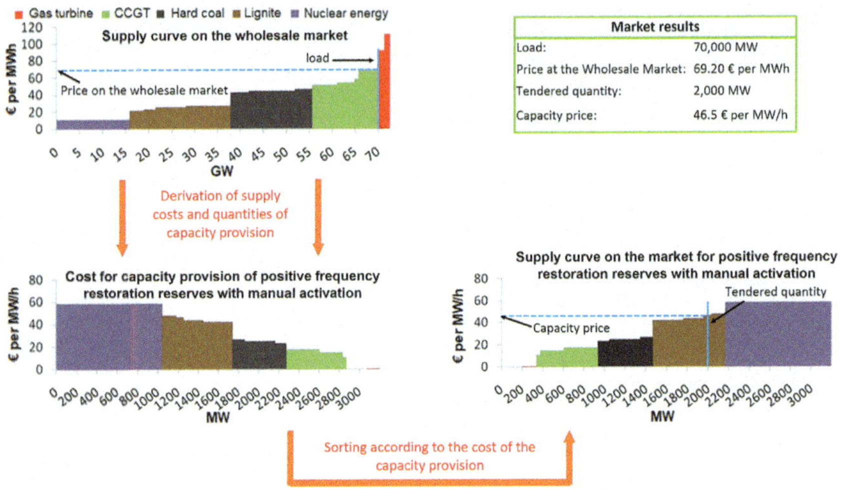

Fig. 3.53 Supply curve derivation for positive mFRR for a demand of 70 GW[48]

In the following step, on the lower right graph of Fig. 3.52, we sort the plants according to the cost of capacity provision. This gives us the merit order on the market for positive balancing power. Knowing the required capacity of 2000 MW, we can calculate the balancing power price of 27.50 €/MWh.

In Fig. 3.53, we can see another example calculation, this time for a demand of 70,000 MW. Now, nearly all power plants are inframarginal and the wholesale price is significantly higher than in the previous example, namely 69.20 €/MWh. Again, we can construct the merit order for positive mFRR by sorting capacity provision costs in an ascending order (lower right side of the figure). If we calculate the equilibrium point for a balancing power demand of 2000 MW as in the previous example, the capacity price in the market for mFRR is now 46.50 €/MWh.

If we compare the two previous examples, we can see that the reserves market has a higher price when the demand is 70 GW, because the opportunity costs of providing this service are higher compared to a demand of 45 GW. In the latter case, there were a lot of hard coal plants very close to the marginal price that could provide balancing power at very low costs. In the case of 70 MW, this is not the case anymore. Almost all power plants are now inframarginal and would therefore suffer a high financial loss if they kept generation capacity free in order to market it as mFRR. This potential financial loss is reflected in the high capacity costs and thus the increased capacity price.

So far, the analysis of balancing power markets focused on general aspects about costs and prices. However, the actual market developments in practice strongly depend on the market design and regulation. In the following, we provide three case studies that show how balancing power markets work in practice. They give

[48] Müsgens et al. (2014).

an overview of the current regulation in Germany and Norway and the respective market developments. On top of that, a third case study shows the current status of the harmonisation of balancing power markets in Europe.

These case studies are particularly important for energy supply companies with a generation focus, as the marketing of balancing energy can represent a significant source of revenue for power plant operators. It is therefore important to understand how these markets are organised in practice and what needs to be done to participate in these markets. In particular, the case studies show how even within a highly harmonised region such as Europe, market design, market participants and therefore potential revenues can vary, especially due to the different physical circumstances in European countries. The case study on European harmonisation provides an insight into what the future design of balancing markets in Europe will look like in the near to medium term. It also sheds light on which parts of the balancing markets are easier to harmonise and where national regulation will still be possible in the future.

Case Study: Balancing power markets in Germany

In Germany, national balancing power markets (FCR, aFRR and mFRR) are operated by the four TSOs (TransnetBW, Amprion, TenneT and 50 Hz). An overview of the regulatory balancing areas is shown in Fig. 3.54.

On the markets for balancing power and energy there are three different products: primary balancing power (FCR), secondary balancing power and energy (aFRR) and tertiary balancing power and energy (mFRR).

FCR is activated automatically in all participating units in case the grid frequency fluctuates above or below defined thresholds values ($49.99\,\text{Hz} < = f < = 50.01\,\text{Hz}$). The primary reserve providers measure the grid frequency independently at the point of generation or consumption and react immediately to the change in grid frequency by increasing or decreasing their power generation of consumption. aFRR is activated by the operating TSO based on the frequency deviation in specific time intervals. All participating suppliers in the secondary reserve are connected via a communication link with the control room of the respective TSO and exchange data in real time. Activation takes place according to the merit order of the energy price bids. mFRR is requested and activated manually via the Merit Order List Server (MOLS) by the TSO.

The prequalification process is carried out by the TSO in whose zone the electrical energy unit is located. The prequalification requirements differ depending on the form of balancing power offered, but in principle there are some uniform regulations that affect all three flexibility products and can in turn be broken down into four central parts:

- IT requirements and rules for data exchange with the TSO.
- Performance of the real-time operation, where according to a standardised test, the compliance with the requirements regarding the product characteristics (FCR, aFRR, mFRR) is checked.

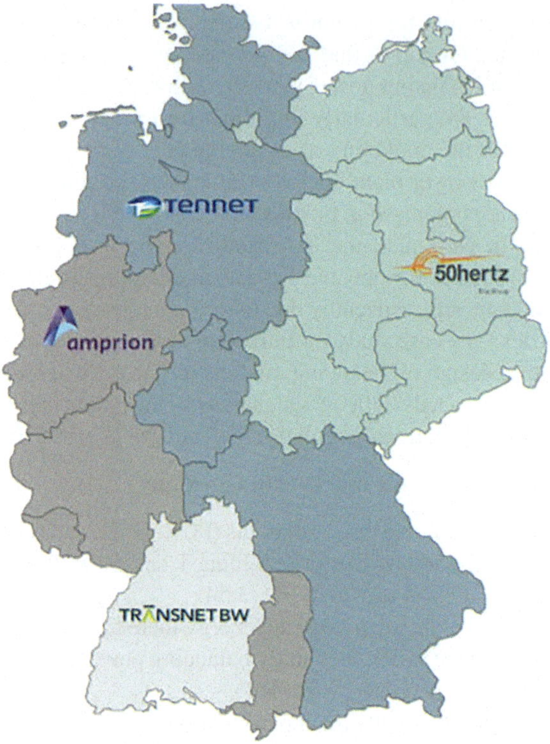

Fig. 3.54 Regulatory balancing areas and the responsive TSOs in Germany[49]

- Control system test, which is carried out at the level of the pool for balancing power.
- Organisational requirements.

Balancing power can be provided either by single units or by pools. A pool is the grouping of technical units with a prequalification for balancing power. The formation of these pools is possible, but subject to more specific prequalification rules than of single units. Also, it makes a difference whether the pool has only one grid connection point or several grid connection points.

All prequalified units can participate in the auctions for balancing power. The transmission system operators maintain their own platform/website on which the auctions take place.[50]

For the primary balancing power (FCR) there is only a capacity market. For secondary (aFRR) and tertiary (mFRR) balancing power, in addition to the capacity

[49] EnBW (2023).

[50] Regelleistung.net (2023b, 2023c, 2023d, 2023a).

Table 3.11 Auction characteristics of German balancing power markets[51]

Market	FCR—capacity market	aFRR—capacity market	aFRR—energy arket	mFRR—capacity market	mFRR—energy market
Tender issue	Daily	Daily	Daily	Daily	Daily
Gate opening time	d−14, 08:00	d−7, 10:00	d−1, 12:00	d−7, 10:00	d−1, 12:00
Gate closure time	d−1, 08:00	d−1, 09:00	d−0, h-25min	d−1, 10:00	d−0, h-25min
Market time unit	6 blocks à 4 h	6 blocks à 4 h	96 blocks à 15 min	6 blocks à 4 h	96 blocks à 15 min
Publication of Results	d−1, 09:30	d−1, 09:30	Depending on PICASSO[52]	d−1, 10:30	Depending on MARI[52]

market, there is also an energy market in which auctions take place and bids can be submitted. For the provision of aFRR and mFRR, there is an obligation to submit a further bid for the energy market after a bid on the capacity market has been placed. However, it is also possible to submit a bid exclusively on the energy market independently of the capacity market.

On the capacity market, the TSOs procure enough capacity to ensure that sufficient reserves are available in the individual periods. The amount held in reserve corresponds to the power that can potentially be called by the TSOs, i.e. what is necessary to keep the system stable. This amount varies depending on the balancing power product and is adjusted each quarter by the TSOs based on statistical calculations. It serves as a safeguard so that there is always enough capacity available.

On the energy market, the order in which the balancing power is called is determined according to a merit order list. The called quantity corresponds to the required amount of balancing energy.

Table 3.11 shows the auction characteristics of the German balancing power markets. Capacity markets (aFRR and mFRR) are organised as 4-h-products. This means that a bid on the capacity market for aFRR or mFRR covers a period of 4 hours. In contrast, the energy markets are 15-min products, i.e. a bid on the energy market is only valid for a single 15-min interval. Capacity markets close the day before delivery, while energy markets close just a few minutes before delivery

Bids on the FCR market must contain

- Market time unit.
- Offered power [MW] in both directions.
- Capacity price [€/MW].
- Divisibility of the bid.

[51] Source: Own illustration based on Regelleistung.net (2023b, 2023c, 2023d).

[52] PICASSO and MARI are the names of European projects for the harmonisation of the balancing energy markets. These will be explained in case study 3.

Bids on the aFRR and mFRR markets must contain

- Market time unit.
- Offered power [MW] and direction (positive/upwards or negative/downwards).
- Capacity price [€/MW] (balancing capacity auction).
- Energy price [€/MWh] (balancing energy auction).
- Divisibility of the bid.

Table 3.12 summarises the products on the German balancing power market and the possible range of bidding parameters. Table provides the values and conditions of the aFRR and mFRR products, respectively. Conditions may change after the transition from the national energy markets to the European platforms PICASSO (aFRR) and MARI (mFRR), which will be explained in case study 3. Apart from these, bids may include further constraints. On the mFRR energy market, block bids are allowed. This can, for instance be a block bid with a technical link for 2 consecutive market time units or a block bid with a conditional link for 3 consecutive market time units.

To minimise overall economic cost, the scoring rules in all markets are generally based on merit orders, i.e. the cheapest bids (up to the required amount) win the auction. The costs are fully covered by the TSOs who perform the tendering. They can later pass on the costs to consumers.

On the capacity markets, the clearing of the auction is operated by the TSOs. On the energy market a Common Merit Order List (CMOL) is built on the PICASSO (aFRR) and MARI (mFRR) platforms since June (PICASSO) and October (MARI) 2022, which are both operated by ENTSO-E.[53] More details of this will be presented in the case study of the European harmonisation later.

Remuneration on the FCR market is based on marginal pricing ("pay-as-cleared"). To compensate a bid, the marginal capacity price (in €/MW) is multiplied with the bid's offered capacity (in MW) and the length of the respective market time unit. Settlement is handled by the TSOs.

For aFRR and mFRR, remuneration of the capacity price is "pay-as-bid", i.e. all bidders receive remuneration based on their specific bid price. The remuneration is calculated by multiplying the offered capacity price (in €/MW) with the offered capacity (in MW) and the length of the market time unit. The remuneration of electricity provided is based on the energy price and is also paid "pay-as-bid". Again, remuneration is calculated by multiplying the offered energy price (in €/MWh) with the actual amount of energy provided (in MWh).[54] The settlement is handled by the TSOs.

If a balancing reserve provider violates its obligations in the modalities, prequalification terms, or the framework agreement with the connecting TSO during the provision or supply of balancing services, the connecting TSO has the right to reduce the remuneration. If the balancing reserve provider does not provide the connecting

[53] European Network of Transmission System Operators for Electricity, see www.entsoe.eu.

[54] This will change according to European harmonisation, where settlement will be done according to marginal pricing and "pay-as-cleared".

Table 3.12 Characteristics of the products on German balancing power markets[55]

Market	FCR—capacity market	aFRR—capacity market	aFRR—energy market	mFRR—capacity market	mFRR—energy market
Minimum size [MW; MWh]	1	1 or 5*	1	1 or 5*	1
Maximum size [MW; MWh]	9999 or 25**	9999	9999	9999 or 25**	9999 or 25**
Size increment [MW; MWh]	1	1	1	1	1
Divisibility	Fully divisible, divisible, or indivisible	Fully divisible	Fully divisible	Fully divisible, divisible, or indivisible	Fully divisible, divisible, or indivisible
Minimum bid price [€/MW(h)]	0.01	0.01	0.01	0.01	0.01
Maximum bid price [€/MW(h)]	9999	9999	9999	9999	9999
Bid price increment [€/MW(h)]	0.01	0.01	0.01	0.01	0.01
Type of activation	Automatic	Automatic	Automatic	Manual	Manual
Full activation time [min]	0.5 (30s)	5	5	12.5	12.5
Minimum duration [min]	No minimum	No minimum	No minimum	5	5

* Minimum bid size 5 MW if multiple bids are submitted in the same balancing zone of a TSO. If only one bid is submitted in a specific balancing zone, the bid size has a minimum of 1 MW
** Indivisible bids have a maximum bid size of 25 MW, fully divisible bids of 9999 MW

TSO with the data in due time, the service is considered as not provided in the respective period. In the event of repeated violations of the correct delivery of balancing energy, the provider is warned and henceforth participates in the balancing power market on probation. If the provider continues to fail to deliver balancing energy properly, the provider loses its prequalification.

Figure 3.55 illustrates the development of the quantities of balancing power held in reserve as well as balancing energy that was actually called.[56] Also, the remunerated

[55] Source: Own illustration based on Regelleistung.net (2022).

[56] Bundesnetzagentur (2023).

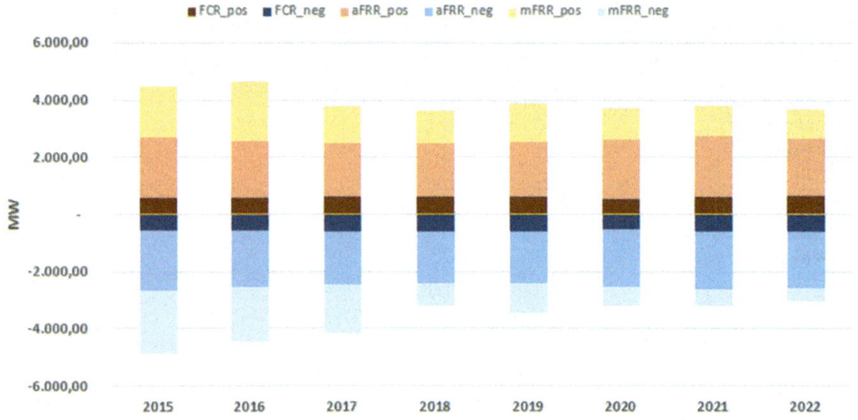

Fig. 3.55 Mean held balancing capacity per quarter-hour of FCR, aFRR and mFRR in Germany (10/2015–09/2022)[58]

prices are illustrated. All values were normalised to 15 min intervals if the original data from the source was not available in 15 min intervals.[57]

Regarding the capacities held in reserve, the total quantity of all types of balancing power has been decreasing since 2015. FCR and aFRR (positive and negative) remain comparatively constant, but significantly less negative and positive mFRR is held in reserve. The reasons for the decreasing amounts of reserved capacity are not entirely clear. Possibly the demand for balancing energy to stabilise the grid frequency is generally decreasing, which is why the TSOs are tendering lower quantities.

The quantities of balancing energy requested have also been declining since 2015, see Fig. 3.56. aFRR is relatively and quantitatively the most frequently called in both directions, while mFRR is almost no longer called in either direction. Usually, the activation of the aFRR is already sufficient to stabilise the grid frequency.

Figure 3.57 shows the mean capacity prices in €/MW per quarter-hour from 2015 to 2022. In recent years, the earnings for offered capacity on the mFRR market have constantly declined. Capacity prices for aFRR have also decreased in positive as well as negative direction. However, they have stabilised on a level of around 2 €/MW per quarter-hour. The capacity prices for FCR have been increasing since 2020 again. The reasons for these price shifts are not entirely clear. One possible explanation is that the mandatory symmetrical (positive and negative) provision of FRC leads to higher opportunity cost, because power plants must keep capacity in both directions free. With increasing electricity prices in 2021 and 2022, this lead to significantly higher opportunity cost for spinning reserve. Power plants providing

[57] This scaling was done to make the values of the respective products comparable, since, e.g. FCR is traded in time periods of four hours on the capacity market and the conversion from aFRR and mFRR to market time units of 15 min happened between July and October 2022.

[58] Source: Own illustration based on data from smard.de.

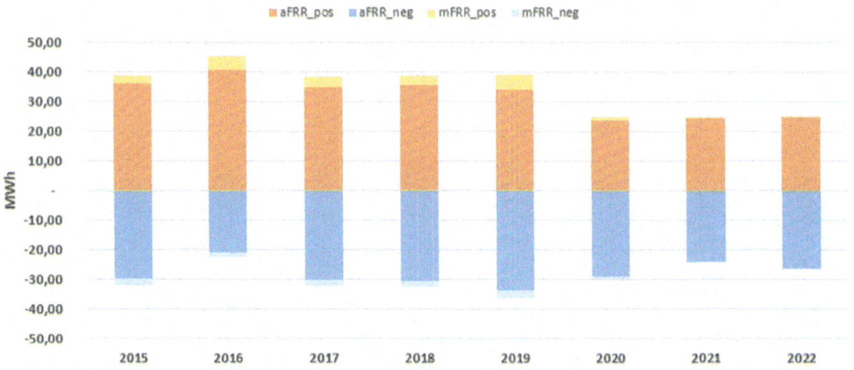

Fig. 3.56 Mean called balancing energy per quarter-hour of aFRR and mFRR in Germany (10/2015–09/2022)[59]

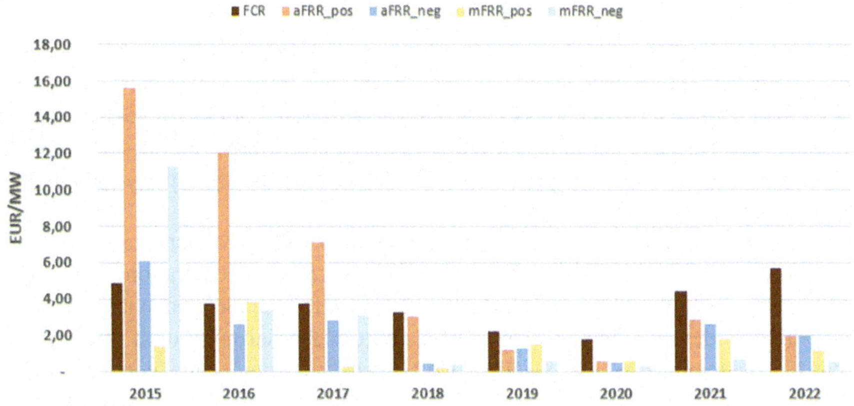

Fig. 3.57 Mean capacity price per quarter-hour of FCR, aFRR and mFRR in Germany (10/2015–09/2022)[60]

aFRR and particularly mFRR, however, do not necessarily have to be spinning and hence have lower opportunity cost.

Figure 3.58 shows the mean energy prices in €/MWh from 2015 to 2022. Tertiary balancing power (mFRR) is rarely called in Germany and hence, energy prices per quarter-hour have been at a very low level in the last eight years. In contrast to that, energy prices for aFRR have been strongly increasing in the recent past. The reason for this price increase for aFRR is not fully clear, but most likely caused by the increasing fuel prices since 2021.

[59] Source: Own illustration based on data from smard.de.

[60] Source: Own illustration based on data from smard.de.

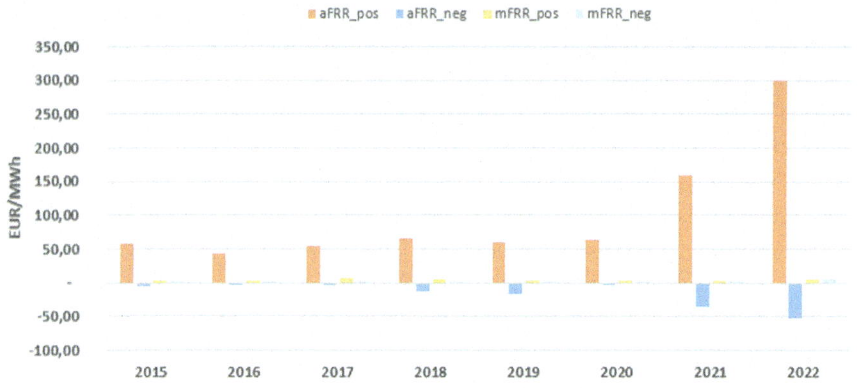

Fig. 3.58 Mean energy price of aFRR and mFRR in Germany (10/2015–09/2022)[61]

Case Study: Balancing power markets in Norway

This case study shows the current state of balancing power markets in Norway. The aim of this case study is to show that even in another European country, the design of the balancing energy markets can be largely different from that in Germany. The products are partially harmonised, but as we will show, there are also completely different products that require different characteristics from the power plants that want to participate in these markets.

The principles and procedures for balancing the Nordic grid system are following the Nordic Balancing Model from the "Nordic System Operation Agreement (SOA)",[62] which has been implemented for and by the Scandinavian countries Norway, Sweden, Finland and Denmark as part of the harmonisation of the European electricity markets. The Swedish TSO (*Svenska Kraftnät*), the Norwegian TSO (*Statnett*) the Danish TSO (*Energinet*) and the Finnish TSO (*Fingrid*) are the main operators of the platforms as well as the responsible entities for balancing frequency. An overview of the different bidding zones of the *Nordics* is shown in Fig. 3.59.

On the capacity markets for balancing power there are multiple different products tradeable:

- **Fast Frequency Reserve**: The objective of the Fast Frequency Reserve (FFR) is to assist the Frequency Containment Process (FCP) during times of low system inertia. This works such that after a sudden imbalance the frequency change can be successfully stopped before the instantaneous frequency deviation would have reached the maximum instantaneous frequency deviation. FFR is only procured as upregulation if the system inertia is too low and the size of the reference event is so large that the frequency containment reserve for disturbances (FCR-D) alone is not able to contain the frequency. In the "FFR profil" product, the flexibility provider offers a fixed capacity for a certain number of hours (usually during

[61] Source: Own illustration based on data from smard.de.

[62] Affärsverket svenska kraftnät; Energienet; Fingrid Oyj; Kraftnät Atland Ab; Statnett SF, 2019.

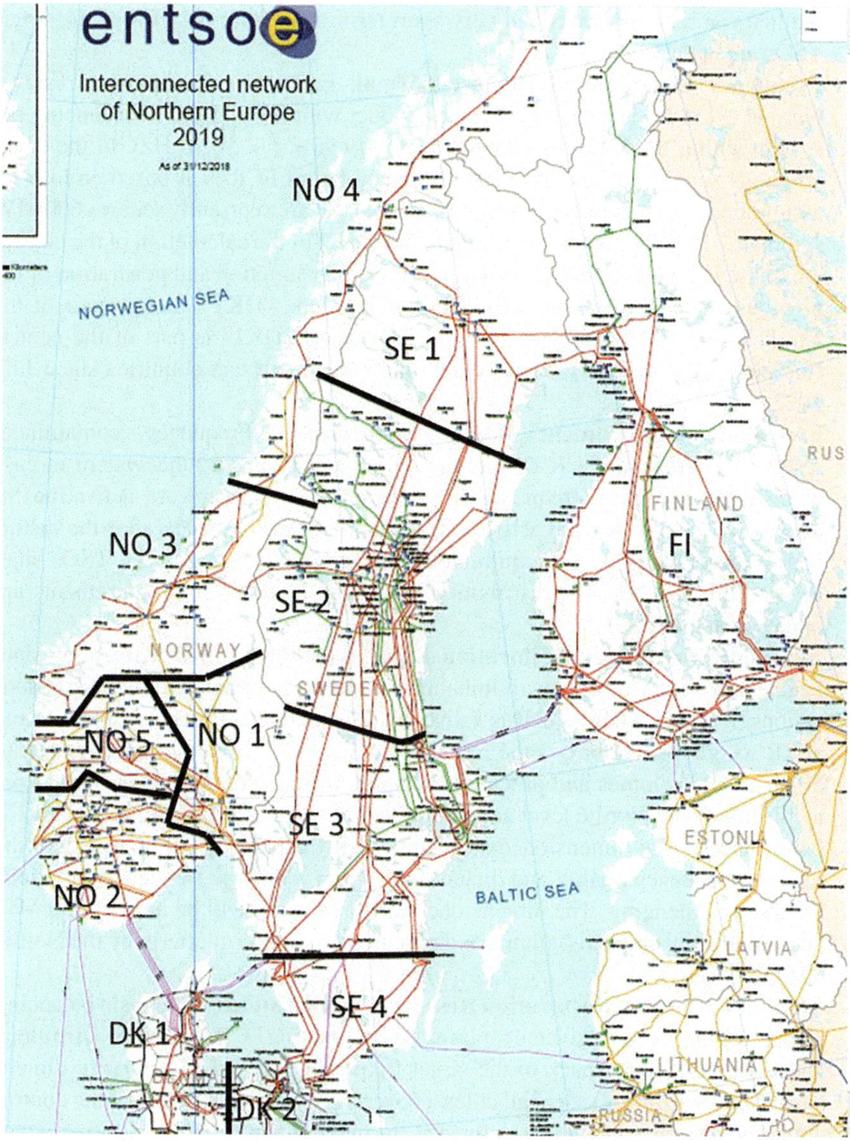

Fig. 3.59 Current Nordic bidding zones[63]

night hours from 00:00 to 09:00 or on weekends) during the season from May to September (as the demand for FFR is particularly high during this phase). With "FFR flex", on the other hand, the provider commits to providing a guaranteed amount of flexibility over a certain number of hours. During this provision period,

[63] ENTSO-E (2019).

Statnett can request capacity at very short term. Shares for the FFR of the single TSOs are updated every year.

- **Frequency Containment Reserve-Normal**: Frequency containment reserve-Normal (FCR-N) is a specific Nordic product with the purpose of balancing the system within the normal frequency band ($49.90 < f < 50.10$ Hz). In the "synchronous zone", the dimensioning of the amount of FCR-N is based on historic assumptions of random load variation. The Nordic area currently secures 600 MW symmetric FCR-N capacity throughout the year. For the calculation of the control area-based quantities for the coming year, the consumption and generation of the previous calendar year are used. The bidding zone "DK1", though part of the Nordics, does not participate in FCR-N, because "DK1" is part of the central European FCR market with Germany and other European countries since July 2022.
- **Frequency Containment Reserve-Disturbance:** Frequency containment reserve-Disturbance (FCR-D) has the purpose of balancing the system in case of disturbances where frequency drops below 49.5 Hz (upward activation) or above 50.5 Hz (downward activation) to stabilise the frequency after the disturbance. Distribution of the requirement for the FCR-D between the TSOs shall be according to the same distribution key as for FCR-N. The requirements are updated each day.
- **Automatic Frequency Restoration Reserve (aFRR):** aFRR activations shall handle short-term variations in imbalance which are not handled by mFRR activations. In hours where aFRR is active, there is an interaction between FCR and aFRR where FCR stabilises the frequency while aFRR brings frequency back to 50.00. aFRR volumes and procurement hours for the Nordic Synchronous Area are decided on a Nordic level and distributed between TSOs by an agreed distribution factor. The dimensioned amount of aFRR capacity shall be based on the targeted frequency quality and includes the hours where the frequency variations are most challenging. The dimensioned aFRR capacity will be at least 300 MW and is currently updated (volume and procurement hours) quarterly by the Nordic TSOs.
- **Manual Frequency Restoration Reserve (mFRR)**: mFRR is the main balancing resource which when activated replaces both remaining FCR and aFRR activations and brings frequency back to the target frequency. The mFRR capacity dimensioned for the control area shall at least cover the reference incident of the control area. The reference incident is defined as the maximum positive or negative power deviation occurring instantaneously between generation and demand.

Table 3.13 shows the shares of balancing power products in the Nordic area for 2022.

For primary balancing power (FCR-N and FCR-D), there exists only a capacity market. For aFRR, until 2022 there were only national capacity markets, but as of the 7th of December 2022, the common Nordic aFRR capacity market went live. This harmonised capacity market comes along with the implementation of cross-zonal capacity reservations between the participating TSOs. In 2024, a common Nordic

Table 3.13 Shares of the balancing products between the Nordic TSOs for 2022[64]

TSO	FCR share (%)	FCR share (MW)	FFR share (%)	FFR share (MW)	aFRR share (%)	aFRR share (MW)
Energinet	2.74	NA	8	24	10	30
Fingrid	19.88	NA	18	54	20	60
Statnett	39.05	NA	39	117	35	105
Svenska Kraftnät	38.33	NA	35	105	35	105

energy market for aFRR shall be launched, which will be adapted to the European platform PICASSO.

For mFRR, there exists a national capacity market (to ensure that enough bids are offered in the energy market) and a common Nordic energy activation market (EAM). The current capacity market conducts weekly and seasonal auctions. The national capacity markets for mFRR shall be transitioned to a new Nordic mFRR market, which will include daily auctions.[65] The current common Nordic energy market for mFRR will also be developed and aligned with the MARI platform in 2024 over the course of European harmonisation.

In the following, the main characteristics of the different balancing power markets in Norway are presented.

FFR (profil & flex):

- Tender issues: seasonal (May–September).
- Market: Capacity Market.
- Gate opening time: variable (in 2022: Jan. 31st).
- Gate closure time: variable (in 2022: Feb. 28th).
- Market time unit: fixed capacity over the season for certain hours with high need, typically night hours and weekends in summer (FFR profil); guaranteed amount of delivery when ordered on request from Statnett (FFR flex).

FCR-N & FCR-D:

- Market: Capacity market.
- Tender issues: daily (d-1 & d-2).
- Gate opening time: not publicly available, published shortly before beginning of the season.
- Gate closure time: d-1 (18:30); d-2 (18:30).
- Market time unit: 24 blocks à 1 h

[64] ENTSO-E (2022); share of mFRR was not available.

[65] For the latest information on the status of the project, please visit https://nordicbalancingmodel.net/roadmap-and-projects/nordic-mfrr-capacity-market/

aFRR:

- Market: Capacity Market.[66]
- Tender issue: daily (d-1).
- Gate opening time: d-7 (00:00).
- Gate closure time: d-1 (07:30).
- Market time unit: 24 blocks à 1 h

mFRR:

- Markets: Capacity market (National), Energy market (Nordics).
- Tender issue:
  - Capacity: daily.
  - Energy: daily.
- Gate opening time:
  - Capacity market: d−7 (00:00).
  - Energy market: d−7 (00:00).
- Gate closure time:
  - Capacity market: d−1 (07:30).
  - Energy market: h−45 min
- Market time unit:
  - Capacity market: 24 blocks à 1 h
  - Energy market: 96 blocks à 15 min

The scoring rules for FFR (profil & flex), FCR-N and FCR-D are merit order based, i.e. the cheapest bids up to the required amount win the auction. For the aFRR capacity market, there is also a merit order list, where the cheapest bids up to the required amount win the auction. Until Go-Live of the harmonised Nordic aFRR energy market or connection to the European PICASSO platform, aFRR is activated *pro-rata* by the TSO, meaning that all contracted aFRR providers will be activated for identical amounts of aFRR whenever there is a frequency deviation within the operated zone. On the mFRR capacity and energy market, there is also a merit order, where the cheapest bids up to the required amount win the auction.

Regarding renumeration, for FFR (profil & flex), there is marginal pricing ("pay-as-cleared"). The settlement is equal to the accepted capacity bid volume (in MW) multiplied by the capacity clearing price (in €/MW). The activation of energy is reimbursed, too, but the methodology for calculating the reimbursement is not clear yet. For FCR-N and FCR-D, there is also marginal pricing on the capacity markets ("pay-as-cleared"). The settlement is equal to the accepted capacity bid volume (in MW) multiplied by the capacity clearing price (in €/MW). Delivered energy for

66 There is currently no national or common Nordic energy market for aFRR, but this is planned within the PICASSO project of the European Union.

FCR-N is valued at the mFRR price in the imbalance settlement. Delivered energy for FCR-D is priced at 0 in the imbalance settlement.

On the aFRR capacity market, there is also marginal pricing ("pay-as-cleared"). The settlement is equal to the accepted capacity bid volume (in MW) multiplied by the clearing price (in €/MW). The price settlement for the activation of energy (until transition to the PICASSO market platform) is currently set equal to the energy price for activated mFRR. On the mFRR capacity market as well as on the energy market prices are set by marginal pricing ("pay-as-cleared"). The settlement is equal to the accepted/activated capacity/energy bid volume (in MW/MWh) multiplied by respective capacity/energy clearing price (in €/MW or €/MWh). If required, the TSO can also activate additional bids according to the "pay-as-bid" principle. As mentioned above, the energy price for activated mFRR is also the price for activated aFRR.

Case Study: Harmonisation of balancing power markets in Europe

In this section, we describe the current status of European harmonisation of flexibility markets. This consists of several projects, which are coordinated by ENTSO-E as parent organisation of European TSOs. The goal of these projects is the same: to create an integrated European market for ancillary services. At this point in time (November 2023), the projects have not yet been fully implemented and the implementation status of the projects varies. The desired final realisation of the implementation is therefore presented below.

Following the Electricity Balancing Guideline by **ENTSO-E**,[67] the relevant European markets for flexibility are the balancing reserve markets as well as wholesale markets for electricity. Moreover, a draft for harmonised regulations for the ancillary service of Demand Response has been published by **ACER** in December 2022.

Balancing markets for capacity reserve and energy provision exist in most of the European Countries. In some countries, there are only energy markets or mandatory capacity reserve conditions for certain market participants. In other countries, market-based mechanisms are already used to provide balancing power (capacity and energy) in case of frequency deviations.

So far, the markets for FCR, aFRR (project **PICASSO**), mFRR (project **MARI**) as well as RR[68] (project **TERRE**) are already in advanced stage regarding the participation of European countries and the synchronisation of common regulations. Those four balancing markets cover the most important system services that are needed to stabilise the grid frequency. By the end of 2022, France, Germany, Slovenia, Austria, Switzerland, Belgium, Netherlands, Czech Republic and Western-Denmark were project members and operated in the common FCR market (Figure 3.60). More countries shall be included in the next years.

---

[67] Current consolidated version: 13.03.2021.

[68] RR is the abbreviation for replacement reserve. This is also known as quaternary reserve and is a reserve in the power grid that helps to keep the grid frequency stable.

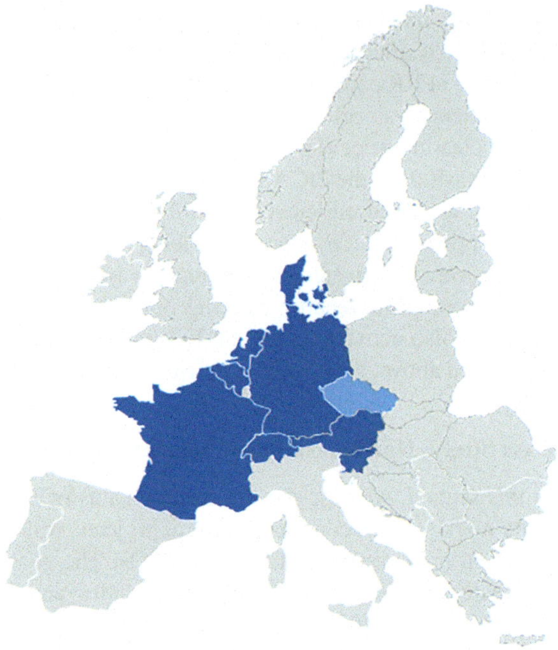

Fig. 3.60 FCR cooperation[69]

For FCR, there is only a capacity market, which is already in a core-European cooperation. Currently, there is no separate energy market planned. Some countries may have special regulations regarding the reserve and provision of FCR or equivalent products. FCR is automatically activated by a frequency-controlled signal and has a full activation time of 30 seconds. The primary reserve provider measures the grid frequency independently at the point of generation or consumption and reacts immediately to the change in the grid frequency.

Figure 3.61 shows the current participants in the PICASSO project for the harmonisation of aFRR. Austria, Belgium, Croatia, Czech Republic, Denmark, Finland, France, Germany, Hungary, Italy, Netherlands, Norway, Poland, Portugal, Romania, Slovakia, Slovenia, Spain and Sweden are currently (December 2022) already members and participate jointly in the harmonised energy market for aFRR. Bulgaria, Greece and Switzerland are currently still observers and may also participate in PICASSO in the future.

Figure 3.62 shows the current participants in the MARI project for the harmonisation for mFRR. Austria, Belgium, Croatia, Czech Republic, Denmark, Estonia, Finland, France, Germany, Greece, Hungary, Italy, Latvia, Lithuania, Netherlands, Norway, Poland, Portugal, Romania, Slovakia, Slovenia, Spain, Sweden, Serbia, Switzerland and UK are currently (December 2022) already members on the European MARI platform. Bulgaria, Ireland and Norther-Ireland are currently observers and may also participate in MARI in the future.

[69] ENTSO-E, Frequency Containment Reserves (FCR) (2023).

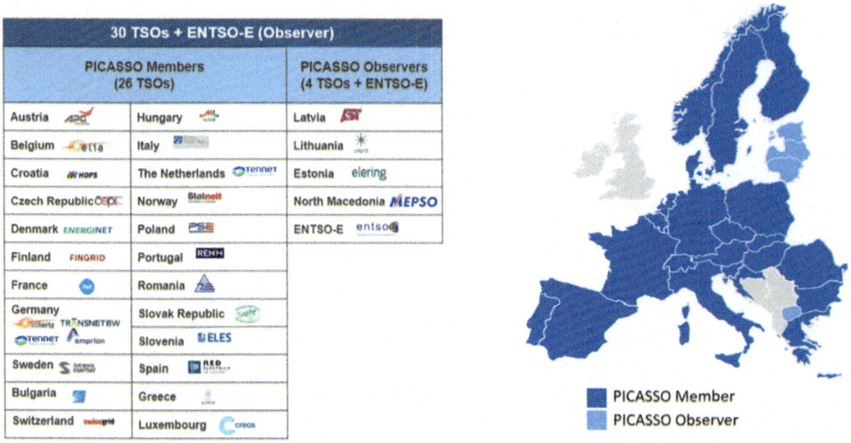

30 TSOs + ENTSO-E (Observer)		
PICASSO Members (26 TSOs)		PICASSO Observers (4 TSOs + ENTSO-E)
Austria	Hungary	Latvia
Belgium	Italy	Lithuania
Croatia	The Netherlands	Estonia
Czech Republic	Norway	North Macedonia
Denmark	Poland	ENTSO-E
Finland	Portugal	
France	Romania	
Germany	Slovak Republic	
	Slovenia	
Sweden	Spain	
Bulgaria	Greece	
Switzerland	Luxembourg	

Fig. 3.61 Current participants in the PICASSO project for aFRR[70]

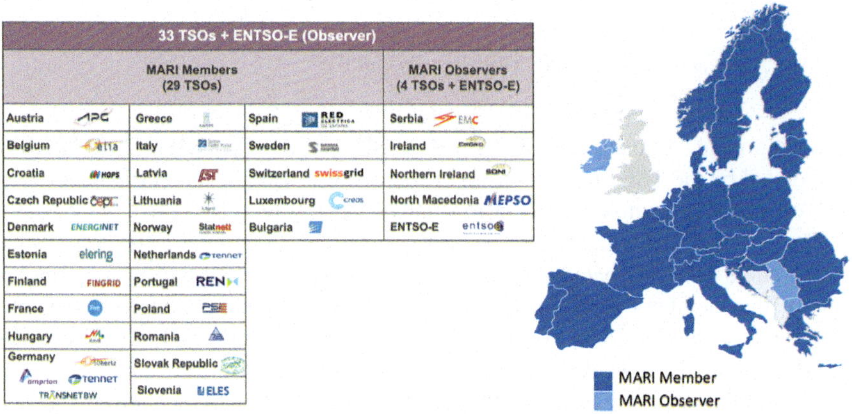

33 TSOs + ENTSO-E (Observer)			
MARI Members (29 TSOs)			MARI Observers (4 TSOs + ENTSO-E)
Austria	Greece	Spain	Serbia
Belgium	Italy	Sweden	Ireland
Croatia	Latvia	Switzerland	Northern Ireland
Czech Republic	Lithuania	Luxembourg	North Macedonia
Denmark	Norway	Bulgaria	ENTSO-E
Estonia	Netherlands		
Finland	Portugal		
France	Poland		
Hungary	Romania		
Germany	Slovak Republic		
	Slovenia		

Fig. 3.62 Current participants in the MARI project for mFRR[71]

Figure 3.63 shows the current participants in the TERRE project for the harmonisation of RR. France, UK, Italy, Portugal, Spain, Switzerland, Czech Republic and Poland are currently (December 2022) already members of TERRE and participate in the European platform for RR. Bulgaria, Hungary, Romania, Germany and Norway are observers.

For aFRR, mFRR and RR, there are energy markets and—depending on the country—may also be capacity markets. The market platform for balancing energy is operated by ENTSO-E together with the national TSOs. The TSOs are responsible for capacity reserve and the electricity flows in the transmission grid of their own

[70] ENTSO-E, (PICASSO) (2023).

[71] ENTSO-E, (MARI) (2023).

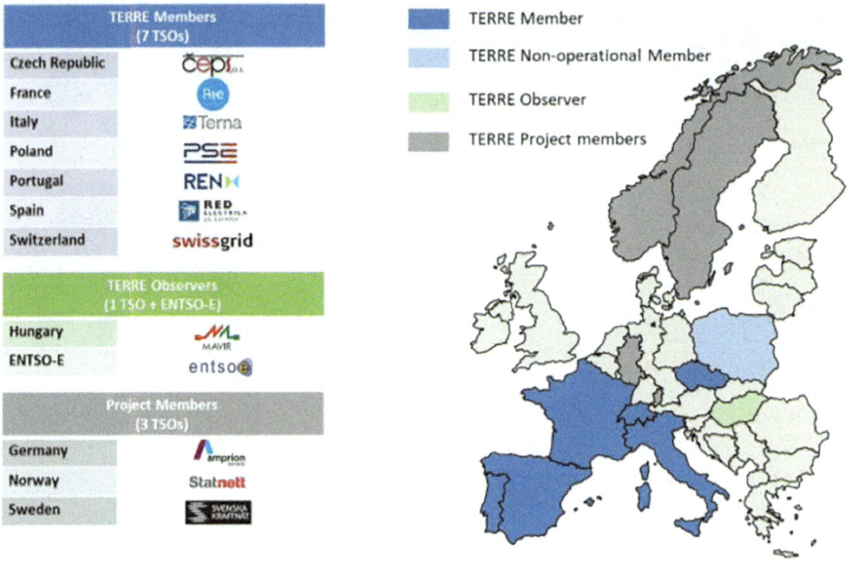

Fig. 3.63 Current participants at the TERRE project for RR[72]

balancing zone. Activation of aFRR (PICASSO), mFRR (MARI) or RR (TERRE) is operated by the TSOs by measuring the deviations from the target frequency. The type of activation (automatically or manually) depends on the product.

The national TSOs in their respective control areas continue to be responsible for the auctions on the balancing market for capacity. For FCR, the capacity market bids of the members participating in the joint FCR market are forwarded to the FCR platform by the TSOs. A European capacity market for aFRR, mFRR and RR coordinated by the ENTSO-E is not planned so far.

The regulated energy market at the European level is created on platforms named after the projects PICASSO (aFRR), MARI (mFRR) and TERRE (RR). The TSOs are responsible for collecting the bids for the balancing energy markets (aFRR, mFRR and RR) and transmit these bids to the European platforms so that they can be compiled in a common European merit order list (CMOL).

Table 3.14 summarises the characteristics of the different European markets for balancing power. There are daily auctions for all products, but both gate opening and gate closure times differ from product to product. Also, the market time unit for the FCR capacity market is four hours, while for the energy markets of aFRR, mFRR and RR, they are only 15 min.

The timelines of the harmonised markets are visualised in Fig. 3.64. Apart from balancing power, this figure also shows the timelines for Intraday markets (ID) and day-ahead markets (DA). Gate closure times (GCT) of RR (TERR), aFRR (MARI)

[72] ENTSO-E, (TERRE) (2023).

Table 3.14 Auction characteristics of harmonised balancing markets in Europe

	FCR	aFRR (PICASSO)	mFRR (MARI)	RR (TERRE)
Tender issue	Daily (d−1)	Daily (d−1)	Daily (d−1)	Daily (d−1)
Market	Capacity market	Energy market	Energy market	Energy market
Gate opening time	Depending on national rules by TSOs	Depending on national rules by TSOs	Depending on national rules by TSOs	Depending on national rules by TSOs
Gate closure time	d−1 (08:00)	h−25 min	h−25 min	h−55 min
Market time unit	6 blocks à 4 h	96 blocks à 15 min	96 blocks à 15 min	96 blocks à 15 min

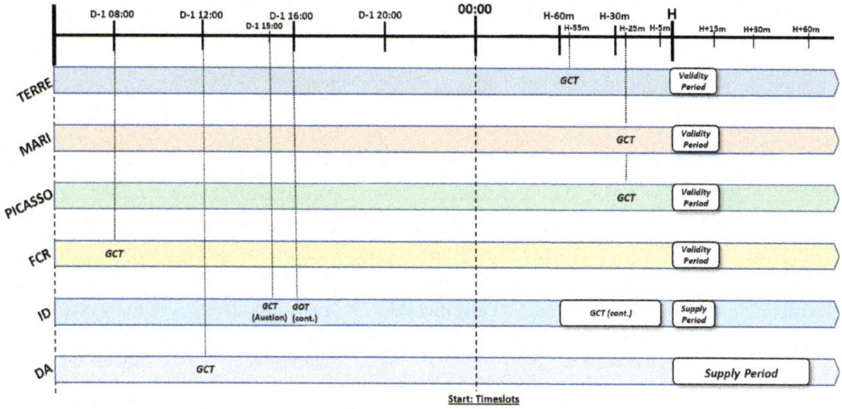

Fig. 3.64 Timeline of energy products in European Harmonisation[73]

and mFRR (PICASSO) are very close to delivery (validity) period and are very similar to continuous intraday trading (ID).

Table 3.15 shows the product definitions on the different harmonised balancing power markets in Europe. In addition to simple bids, complex bids are also possible. This may be exclusive groups, block bids and linking of bids. The specific types of permitted complex bids vary between the different products. For FCR, complex bids or linking of simple bids is not permitted. For aFRR (PICASSO), mFRR (MARI) and RR (TERRE), complex and linked bids. Technical units may also be aggregated to pools.

Figure 3.65 shows the remuneration principle for the harmonised markets, which is always based on the pay-as-cleared principle, i.e. the marginal bid sets the market price (cross-border marginal price—CBMP) paid to all participants. For remuneration on the FCR market, a Common Merit Order List (CMOL) on European Level is calculated. This CMOL considers national constraints (cross-zonal capacities) and

[73] EPEX SPOT (10.2022).

Table 3.15 Product definitions of harmonised balancing markets in Europe

	FCR	PICASSO (aFRR)	MARI (mFRR)	TERRE (RR)
Bid time resolution/validity period [min]	240	15	Scheduled activation (SA)—at the time of scheduling Direct activation (DA)—at any time during the 15 min interval after the scheduled activation time	Defined by balance service provider (BSP) Min.: 15 min Max.: 60 min
Minimum size [MW]	1 or 25 (indivisible bids)	0 (fully divisible bids) or 1	0 (fully divisible bids) or 1	0 (fully divisible bids) or 1
Maximum size [MW]	NA	9999	9999	9999
Increment size [MW]	1	1	1	1
Minimum price [EUR/MWh]	NA	−99,999	−99,999	−99,999
Maximum price [EUR/MWh]	NA	99,999	99,999	99,999
Price granularity [EUR]	0.01	0.01	0.01	0.01
Direction	Symmetric	Pos./Neg	Pos./Neg	Pos./Neg
Divisibility	Fully divisible, divisible or indivisible	Fully divisible	Fully divisible, divisible or indivisible	Fully divisible, divisible or indivisible

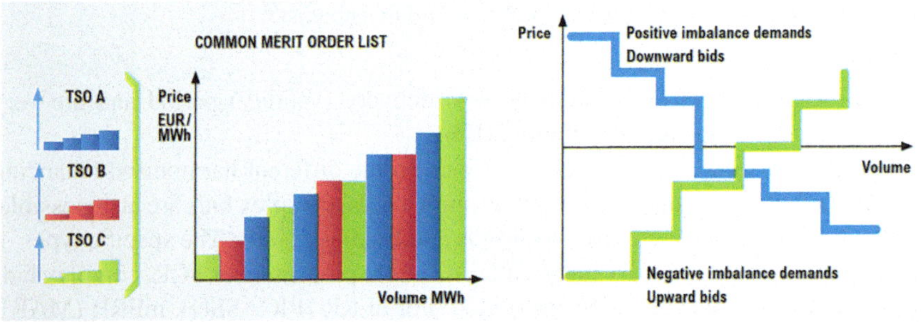

Fig. 3.65 Relation between common merit order list and demand for negative/positive balancing energy[74]

national minimum needs. Winners are the cheapest bids until demand is met. For aFRR (PICASSO), mFRR (MARI) and RR (TERRE), there is also a Common Merit Order List (CMOL) for activation for each product. This also considers national

[74] ENTSO-E (2018).

constraints (cross-zonal capacities) and national minimum needs. Winners are the cheapest bids until demand is met.

For aFRR (PICASSO), there is one CBMP for positive and one for negative energy. This is paid to all providers which are called within the respective validity period of 15 min. In case of problems (e.g. because the bids of the energy market were not sent by the TSOs to ENTSO-E for technical reasons), the respective TSO falls into a national mode and forms its own local merit order list.

Regarding remuneration for mFRR (MARI), there is a separation between scheduled and direct activation. For scheduled activation, two CBMPs are calculated (one for the positive and one for the negative direction) and paid to all participants that are called within the respective validity period.

Since energy delivery in direct activation can last until the next quarter-hour, four CBMPs are calculated for bids of direction activation (one for positive and one negative direction in the current quarter-hour and one for positive and one for negative in the next quarter-hour). The price used to remunerate positive energy for direct activation is the maximum of this CBMP and the CBMP for scheduled activation in the respective quarter-hour. Consequently, the energy delivered in the first quarter-hour of a direct activation may be remunerated differently from the energy delivered in the second quarter-hour, because the CBMP of the scheduled activation in the two quarter-hours may be different.

In the following, an example of the calculation for the remuneration of called mFRR via direct activation is given.[75] In the case of scheduled activation or if direct activation is performed exactly as scheduled activation, i.e. starting at the beginning of one quarter-hour, the amount of energy that is remunerated corresponds to the energy requested in the respective quarter-hour. It is renumerated using the CBMP in the respective quarter-hour. This is shown in Fig. 3.66.

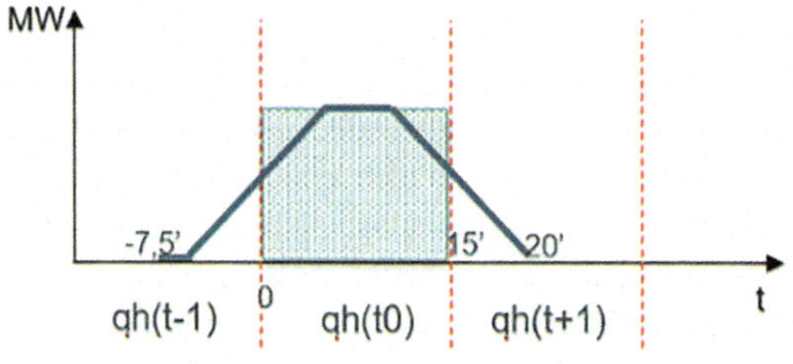

Δt = 15 (no delay)

$$\text{mFRR Energy Requested} = \text{mFRR Requested} * \frac{1}{4} \text{ [MWh]}$$

Fig. 3.66 Remuneration for mFRR (direct activation) without delay[76]

[75] Elia Group (2022).

[76] Elia Group (2022).

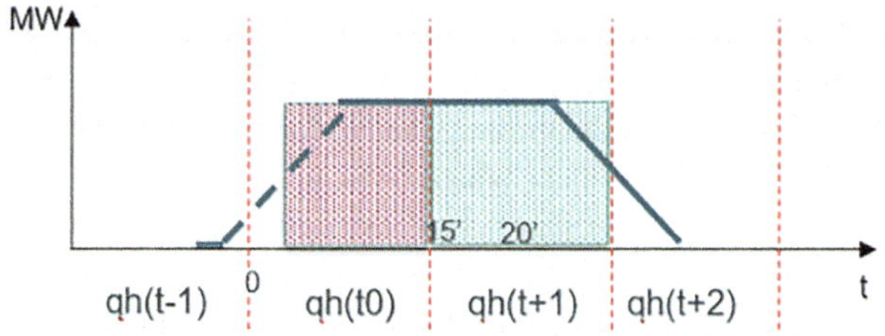

Example: DA request at 2,5' after point of SA for Qh(t0)

For Qh(t0): $\Delta t = 15' - 2{,}5' = 12{,}5$

mFRR Energy Requested

$$= \text{mFRR Requested} * \frac{12{,}5}{15} * \frac{1}{4}\ [\text{MWh}]$$

For Qh(t+1): $\Delta t = 15'$

mFRR Energy Requested

$$= \text{mFRR Requested} * \frac{1}{4}\ [\text{MWh}]$$

Fig. 3.67 Remuneration for mFRR (direct activation) with 2.5 min delay[77]

If direct activation is performed after scheduled activation, i.e. a few seconds or minutes after the beginning of a quarter-hour, the amount of energy that is remunerated corresponds to the actual energy delivered in the concerned quarter-hour qh(t0) plus the energy in the following quarter-hour qh($t + 1$), as if it had been delivered in all 15 min of qh(t1). The former is renumerated using the CBMP in qh(t0), and the latter is renumerated using the CBMP in qh($t + 1$). This is shown in Fig. 3.67.

For RR (TERRE), the settlement rules are also based on uniform pricing using the CBMP. In case of problems, the respective TSO falls into a national mode and forms its own local merit order list.

To summarise, these case studies have shown how different regulation can be even in countries that are similar in many respects. There is a tendency across Europe to harmonise energy markets for all products with the exception of FCR. This means that only the call of balancing energy is standardised. In addition, each TSO can organise capacity auctions to ensure that sufficient capacity is available. However, it may well be that there is no capacity market at all in a country or region and only the energy market can be considered as a source of revenue. This makes it on the one hand less attractive for participants, because capacity reservations are not renumerated.

[77] Elia Group (2022).

On the other hand, it makes participation in these markets more attractive, because it simplifies the opportunity cost calculation. It is therefore extremely important for potential participants in the balancing energy markets to know the exact regulatory details of the region in which their power plant is located.

3.3.5 *Optimisation of Generation Assets*

The previous sections showed that there are many potential markets as well as many potential products to trade the output of generation assets. Profit maximisation requires to take all available markets and products into account. A generation asset is naturally long on some markets, e.g. a power plant can sell output on the following markets:

- Wholesale markets for electricity (forwards, futures, day-ahead and intraday market).
- Real optionality: trade the generation asset's make-or-buy decision.
- Markets for balancing capacity:
  - Positive/negative reserves with manual activation.
  - Positive/negative reserves with automatic activation.
- Markets for balancing energy: If capacity sold on the balancing capacity market is called, the power plant operator receives additional income. These possible additional revenues must be taken into account when submitting the capacity bid, as they influence the value of the bid. If high revenues from balancing energy are expected, the power plant operator may submit a lower capacity price bid.
- Heating market revenues (if applicable).
- Avoided grid usage charges (if applicable).
- Potential new markets due to regulatory changes.

Note that some essential interdependencies between these markets exist:

- Trading power/energy on one market affects the possibility to trade power/energy on other markets. For instance, the balancing power market and the day-ahead market can be both complements and substitutes, and there are other interdependencies between these markets that need to be considered.
- Revenue opportunities on balancing capacity and balancing energy markets are interdependent, as expected revenues on balancing energy markets have an impact on optimal bids on the balancing capacity markets, as explained above.
- Interdependencies between the markets for balancing reserves of different qualities (because a megawatt can be sold only once).

Additionally, a power plant is naturally short on other markets (particularly markets for fuel and CO_2 emissions certificates which are determining the variable production costs). Trades on these markets influence the open position of a generation asset and are thus also part of the portfolio management strategy. For example, selling electricity with a forward contract without buying fuel and CO_2 at the same time leaves the portfolio with an open position. If the portfolio management strategy prefers (or demands) it, this open position can be greatly reduced by trading both electricity as well as fuel and CO_2 at the same time.

Maximising profit contributions in these markets requires sophisticated operations and bidding strategies, deep know how and significant financial resources. Also, various tools are used to support decisions. In particular, numerous forecasts are used, both with regard to prices as well as intermittent generation of renewable energy. For wind and solar power, forecasts of expected generation are needed to determine the actual quantities which can be sold. For prices, forecasts on day-ahead and intraday prices for short-term optimisation, capacity prices for balancing capacities, energy prices and activation probabilities for balancing energy are needed. Activation probabilities need to be forecasted because the probability of being called is not 100 per cent and depends on the energy price bid: The lower the energy price bid, the higher the probability of being called and vice versa.

To maximise the revenues of a generation asset, complex dispatch optimisation models for power plants are used. These models are either developed internally or—more typically—provided by consulting firms. They are sometimes offered as integrated software models and their output contains relevant information for the bidding strategy, e.g. at what price should a certain capacity be offered and on which market.

The complexity of the optimisation problem is illustrated in Fig. 3.68, where a typical dispatch model for a power plant is shown. In light blue are the exogenous inputs and in yellow are the endogenous outputs. These are determined by the mixed integer optimisation problem, which maximises the profit under constraints given by the technical properties of the power plant and characteristics of the external markets.

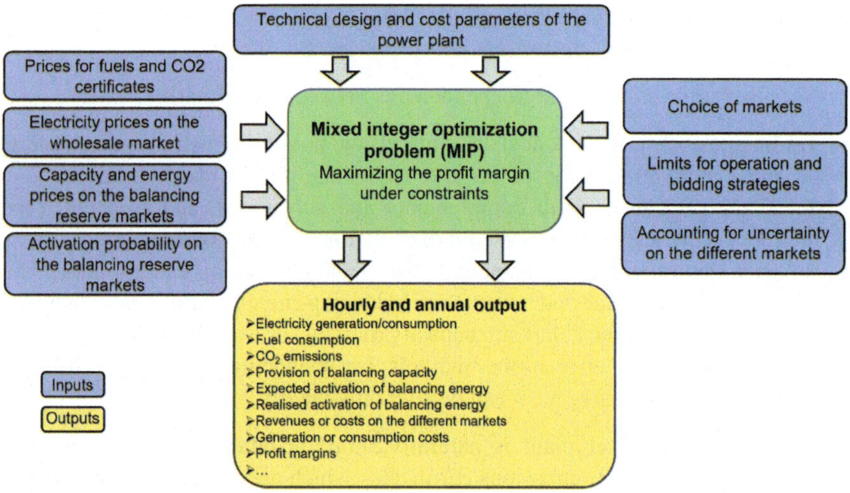

Fig. 3.68 Dispatch optimisation model for a power plant

The most important inputs are the technical design and cost parameters of the power plant (efficiency, start-up costs, minimum up and down times, installed capacity, etc.). Also, future prices of fuels and CO_2 emissions certificates, electricity prices on the wholesale market as well as capacity and energy prices on the balancing reserve markets are required. On top of that, sophisticated models also require activation probabilities on the balancing reserve markets. Some inputs may not be known when a decision is made, and the model is used. They are estimated based on available information. Note that this is the norm in the business world, where decisions need to be taken under incomplete information and uncertainty. Taking good decision under uncertainty is one of the factors driving the success of businesses. As professional means to take uncertainty into account, the operator can include limits for the operation and bidding strategies and may account for uncertainty on the different markets, either with scenarios or with stochastic programming approaches.

Some other inputs depend on the strategy of the power plant operator, i.e. the operator specifies on which markets the plant should sell its capacity. For instance, due to high costs for hardware requirements and the prequalification process, small power plants may decide not to participate in balancing power markets.

Once all the inputs are determined, they can be fed into the mixed integer optimisation model. This model then maximises an objective function, which in this case is the profit margin, while considering all the constraints. There are both technical constraints of the power plant as well as operational constraints such as the fact that the operator may not be active on all of the markets discussed above. The output of this model can be a daily, monthly or annual dispatch of electricity generation in hourly resolution on the different markets, i.e. how much should be offered on the day-ahead market, on the intraday market, on the balancing power market, etc. Additional outputs are, e.g. how much fuel to buy, how many tonnes of CO_2 will be emitted and how many certificates are required to offset these emissions. Furthermore, the expected activation of balancing energy, the revenue or cost on the different markets, generation or consumption costs and profit margins are shown.

Different levels of complexity can be used for these models. In the following, we will show potential outputs of dispatch models starting with a very simple heuristic approach.[78] In this simplified approach, only the day-ahead market is considered and there is no partial load condition for the power plant, i.e. it can be started up and shut down instantly at no cost. A possible output of such a model is shown in Fig. 3.69. The orange dots represent the hourly wholesale electricity prices for three example days. The lower prices occur during the night, and they increase during the day.

The red horizontal dashed line shows the assumed short-term variable cost of electricity production. The model makes a simple comparison between the variable costs and the electricity price and derives a simple dispatch decision: When the hourly price is above the production costs, the plant generates electricity at its maximum capacity. This is shown in the grey bars which represent the hourly power plant

---

[78] This approach is too simplistic in today's competitive markets but serves as a good starting point for didactical reasons.

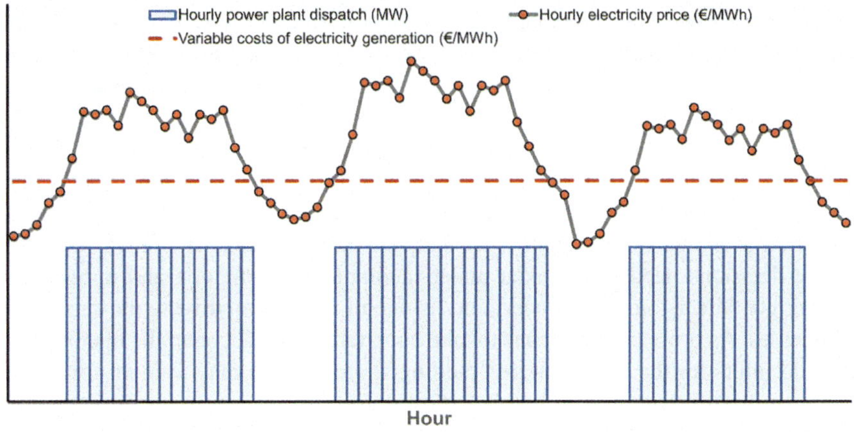

Fig. 3.69 Simple power plant dispatch optimisation

dispatch. When variable costs exceed prices, the power plant is shut down and the electricity production is zero.

In this example, the power plant operator optimises the dispatch of the unit purely depending on the variable cost of electricity generation and hourly electricity prices. This is a simplification that does not consider dynamic effects such as start-up costs, minimum downtimes, and partial load conditions. Also, there is no activity on markets other than the day-ahead market. As a result, this production profile may not be technically feasible or at least not be profit maximising.

Therefore, in a next step, we will consider dynamic effects, and the new dispatch optimisation is shown in Fig. 3.70. Here, the optimal production has changed during the night, and the plant is now generating at minimum load. In this example, the power plant's minimum load is around 40% of its installed capacity. This limitation changes the optimal dispatch: The plant now runs in partial load during the night. In the figure, this is represented by the smaller grey bars during night hours.

Technically, the plant could also be shut down during the night, but start-up costs would be incurred the next morning. However, in our example, the optimisation reveals that after the first start-up around hour six, the plant should remain in operation for all remaining hours—during the day in full load operation and during the night in partial load operation. The economic reason is that negative profit contributions incurred during the night are smaller than the cost incurred with a restart. Of course, this is only profitable when prices are just slightly below variable costs.

There may also be other reasons to keep the plant running when variable costs are higher than the electricity price. One of these is the limitation of the minimum downtime, which prevents premature restarting after the power plant has been shut down. For example, if the power plant has to be shut down for 10 hours after shutdown because the machines need to cool down before restarting, it might be optimal to leave it on the grid.

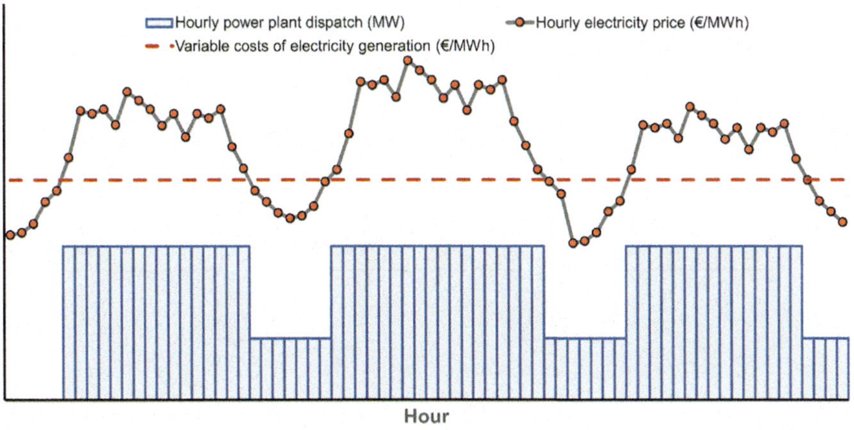

Fig. 3.70 Dynamic power plant dispatch optimisation

Other reasons to keep a unit online are technical conditions which require partial load operation. For instance, in a combined heat and power plant, it may be necessary for the heat system to keep the plant running in cases where there is no backup heat source. It may also be that the backup source is very expensive, and it may therefore make sense to keep the power plant in operation, even if this means facing losses in the electricity market. The same applies to marketing balancing power: If the expected revenues on the balancing power markets exceed losses on the day-ahead market, it makes sense to keep the power plant in operation, even if it generates negative profit contributions on the day-ahead market. While it can thus make perfect sense to operate a plant at prices below costs, generally speaking this is the exception that proves the rule.

In Fig. 3.71 we can see the dispatch with an additional level of complexity. Here, the markets for balancing power are also included in the exemplary optimisation. As a result, the plant provides negative balancing capacity (yellow part of the bars) while running at full load. These additional revenues could even be earned without changing the dispatch profile. When the plant is running at minimum load, positive balancing capacity can be offered, also without changing the operation profile of the plant (light orange part of the bars). In both cases, the plant will generate additional income.

An integrated optimisation on all markets can lead to more significant changes as well. It might, for instance, make sense to keep a certain share of the generation capacity free by lowering the power plant's sales on the day-ahead market below installed capacity to provide positive balancing capacity during the day. Again, this is the complementary part between both markets: Each megawatt can be sold either on the day-ahead or on the balancing power market. This decision is part of the optimisation of an integrated portfolio. The hourly dispatch shown in Fig. 3.71 is closer to reality and is what state-of-the-art models help to determine: How much capacity to offer on each market and then calculate expected revenues for that.

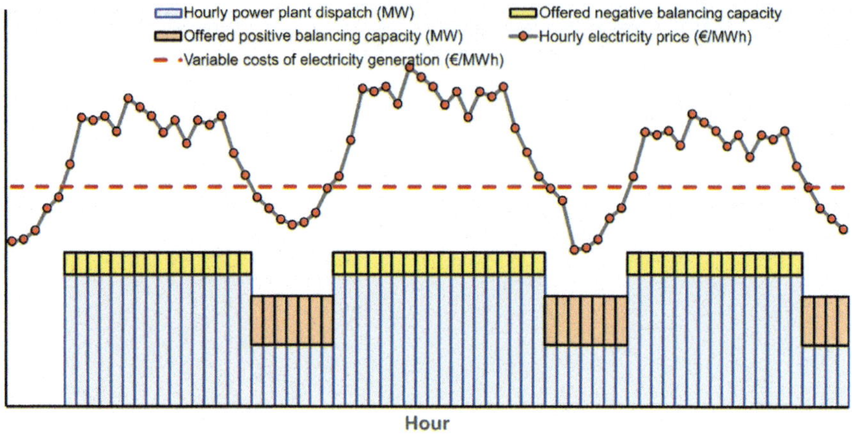

Fig. 3.71 Power plant dispatch optimisation including balancing power

To conclude, the main aspects for marketing flexibility of a power plant (or any other unit that acts on electricity markets) are the following:

- Optimal operation strategies should take into account all markets.
- Due to numerous interdependencies between markets, an integrated optimisation model is preferable.
- The derivation of optimal strategies is complex and requires
  - State-of-the-art forecast and optimisation tools to derive an optimal strategy,
  - Comprehensive data management,
  - Appropriate organisational structures.
- External companies can be used to, e.g. provide
  - Forecasts of day-ahead electricity prices,
  - Forecasts of balancing reserve prices (capacity and energy prices), as well as activation probabilities,
  - Implement marketing and bidding strategies in power plant dispatch optimisation.
- The design of the unit (e.g. load gradients, technical minimum loads, the efficiency depending on the operating point) has a significant impact on potential revenues.

So far, we have focused on how to optimally sell the output of a valuable flexible asset such as a power plant. The question arises as to why flexibility is valued so highly in the system. This is because—as already mentioned—all electricity grids must be operated safely. A key element of a safe grid operation is to maintain a stable frequency, which in turn necessitates supply and demand being always equal. We have already discussed the concept of balancing groups, making sure that on expectation demand equals supply at grid closure on the day before delivery. However, we have also argued that the amount of uncertainty is high in energy systems, for example with regard to the production of wind and solar power or the precise demand profile of a consumer. Deviations between forecasts and realisations for such uncertain parameters need to be compensated on short notice by flexible generation and consumption units. From a regulatory perspective, this requires incentives for power plants (or other units) to provide flexibility. In most modern liberalised energy systems, these incentives are financial payments.

This can be summed up in the following quote from a former minister of economic affairs in Germany regarding the importance of flexibility: "We do not necessarily need more power stations but what we need is flexible capacity. Flexibility is the answer to weather-dependent renewable energy sources. By introducing the electricity market 2.0 we are permitting fair competition between all flexibility options".[79] Here the minister explicitly acknowledges the requirement of flexibility in the system, and policymakers and regulators recognise the need to enable revenue for power plants that provide it. These plants help the system to keep a stable frequency at 50 Hz, and they should be compensated for this service.

3.4 Integrated Portfolio Management

This section summarises the content on portfolio management covered so far and combines it to analyse integrated portfolio management strategies, i.e. companies with both generation assets and retail customers. This will complete and conclude the analysis of portfolio management in energy companies.

Large energy companies often operate generation assets, have retail customers to whom they supply energy and are active on the wholesale market. These are referred to as integrated energy companies. Examples are mostly large municipal utilities, tier-one energy companies and very large industrial customers. As shown in Fig. 3.72, portfolio management is at the centre of their activities, connecting retail clients, wholesale markets and generation assets through an effective and efficient management of the open position.

In this section, we first present advanced financial products that allow for sophisticated portfolio management. After that we discuss how the performance of the different departments involved in trading activities within a company can be measured. We close this section with a case study of a day at the trading floor of an actual energy trading company.

Equation (3.17) defines the open position on an integrated company. For each delivery hour h, it is the purchases (BUY) plus expected generation depending on the variable production costs and the electricity price (GEN) minus the sales (SELL) and consumption (CON), which can be price dependent. Most actors in the energy markets have an open position which is simpler to define, as we have seen, and only large companies have values for all four components of this open position formula. As

---

[79] BMWi (2015).

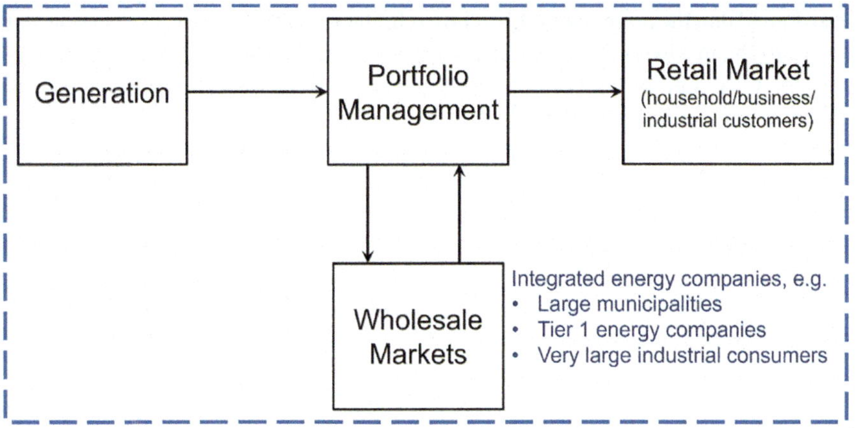

Fig. 3.72 Integrated portfolio management

we have discussed, we need to optimise this open position to achieve the maximum return during the trading period.

$$OP_h = BUY_h + GEN_h(vc_h, p_h) - SELL_h - CON_h(p_h). \qquad (3.17)$$

The open position for a delivery hour h typically varies over time during the trading period. This can happen through transactions, but also through the change in market expectations. For instance, if h is a specific hour of next year, the purchase of a front-year base contract is included in the open position of that hour, as hour h is affected (in this case as part of the BUY term). The expected generation of a power plant depends on the expectation of electricity and fuel prices, i.e. higher fuel prices imply ceteris paribus lower expected generation and vice versa. As we have mentioned before, when registering the schedule on the day before delivery, each balancing group must be balanced—the expected open position for the balancing group must be zero. The BUY and SELL parts of the open position formula are typically known,[80] but it is necessary to decide how much to generate and estimate what the consumption will be.

3.4.1 Advanced Financial Product

The previous sections introduced financial instruments like forwards, futures, options and swaps and explained how to use them in energy markets. In the remainder of this section, we present advanced methods that use these instruments in sophisticated integrated portfolio management strategies. We will show how options can be

[80] An exception are e.g. household customers, whose sales volumes are not 100% known (see Sect. 3.2.1.1).

combined to create complex hedging strategies and how the value of an option (the option premium) can be determined. We will also show how swaps can be used to separate price from supply risk and their financial application as interest rate swaps.

3.4.1.1 Combining Options in Hedging Strategies

The general concept of options has been presented in Sect. 3.3.3. Options can be used to hedge against undesired market price developments of an underlying, which is most commonly a commodity contract for electricity, fuel or emission rights. This concept was extended to spread options, that are related to the payout of physical power plants. However, options offer a lot more opportunities for managing purchase and sales portfolios than that.

The following analysis will go beyond options connected to generation assets and look at how they can be used in an energy utility with retail business, as we described it in Sect. 3.2. In other words, we will look at the possibilities and risks associated with retail business and how options can be used to manage them.

As we have mentioned, there are various possibilities to balance a forward sale that has just been made. If a certain amount of energy was sold to an industrial customer with delivery at some point in the future, there are several options to procure the energy:

1. Procure instantly on the forward market (or by using a swap). This is, as elaborated several times before, a back-to-back procurement, with no risk and no chance. The open position is immediately minimised.
2. Procure on the spot market. This maximises both risk and chance.
3. Hedge with options.
4. Hedge with a zero-cost collar (this will be explained later).
5. Implement a combination of different measures.

To assess the risks associated with each of these strategies, their payout profiles will be explained in the following.

$$\pi = F_{\text{Sale}} - F_{\text{Purchase}}. \tag{3.18}$$

Figure 3.73 shows an exemplary payout profile for the instant closure of the open position (strategy 1 in the list above). In this and all subsequent diagrams, the payout is shown as a function of the spot price at maturity S_T. Immediately after the sale, the sold amount of energy is bought on the future market. In this example, the price for the forward sale F_{Sale} is 55.80 €/MWh and the price for the immediate purchase F_{Purchase} is 55.75 €/MWh. The payout is the difference between the sale price F_{Sale} and the purchase price F_{Purchase}, as shown in Eq. (3.18). The payout is independent of the price on the spot market S_T, because this strategy does not involve a purchase on the spot market.

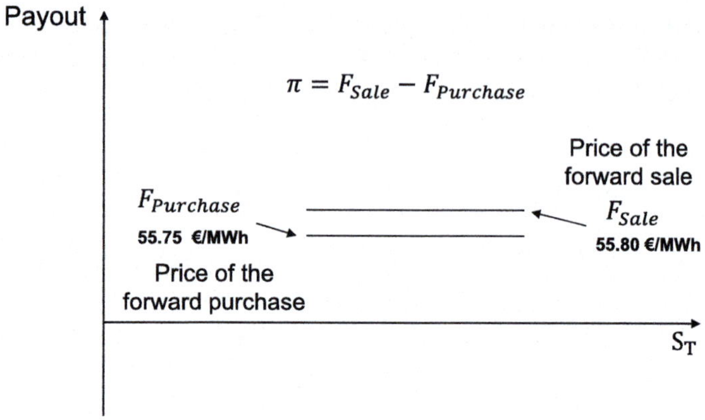

Fig. 3.73 Instant coverage at the futures markets

$$\pi = F_{\text{Sale}} - S_T. \tag{3.19}$$

Figure 3.74 shows the payout profile for closing the open position exclusively on the spot market (strategy 2 in the list above). The figure shows both the chances and the risks this strategy implies. Again, the retail sale is assumed to take place at a price of $F_{\text{Sale}} = 55.80$ €/MWh and there is no immediate coverage of the sold energy on the future market in this strategy. Instead, procurement takes place later on the spot market. In Fig. 3.74, the spot prices can range from zero to very high values, and the payout is the price of the forward sale minus the spot price at maturity, as Eq. (3.19) shows.

If the spot price at maturity S_T is zero, profit contribution would be huge: A commodity was sold for 55.80 €/MWh as a forward and is now being bought for free. That is obviously a theoretical best-case scenario. If S_T increases, the payout

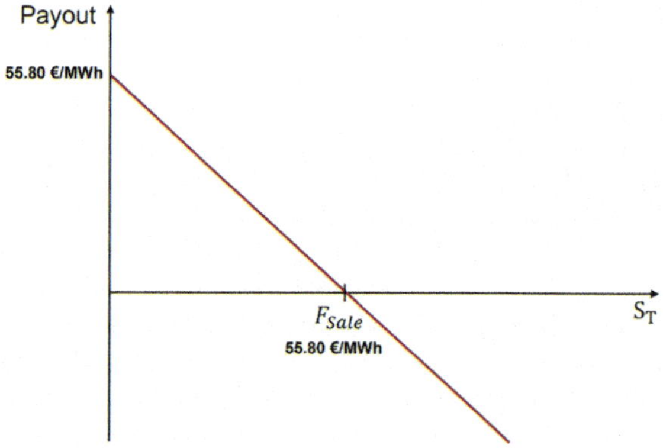

Fig. 3.74 Full coverage on the spot market

starts to fall until it reaches zero at a spot price of 55.80 €/MWh. In this case, the energy would be bought on the spot market at the same price it was sold. Higher spot prices lead to losses. Of course, not all spot prices have the same probability of occurring. Nevertheless, it is worth noting that such a strategy can lead to huge profits, but also to huge losses.

In the third strategy, the forward sale is hedged with a call option. Again, a forward is sold for 55.80 €/MWh, this time together with an immediate buy of an option. Moreover, this strategy includes a purchase on the spot market. This is because the call option is purely a financial hedging instrument and does not involve actual delivery. Physical delivery to the customer thus requires a purchase on the spot market. The strategy's combined payout profile (long call option, the forward sale, and spot market procurement) is shown in Fig. 3.75. The forward sale price F_{Sale} is again assumed to be 55.80 €/MWh, the strike price of the option is assumed to be 65.00 €/MWh and the option premium c is assumed to be 2.60 €/MWh.

$$\pi = F_{\text{Sale}} - S_T + \max\{(S_T - K); 0\} - c. \tag{3.20}$$

The total payout of strategy 3 is simply the sum of the payout of the spot market procurement strategy (strategy 2) plus the payout of the long call option and is given by Eq. (3.20). The corresponding payout profile is shown in Fig. 3.76 (bold red line). In this strategy, there is a positive profit if the spot price is low. However, it is smaller compared to strategy 2 because the option premium must be paid. This can be seen by the bold line being below the dashed line. As long as the spot price is below the strike price K, the payout is equal to the payout of strategy 2 minus the option premium c. At prices above K, the purchase of the option becomes advantageous. The option will be exercised and will result in a positive payout. This payout offsets the loss

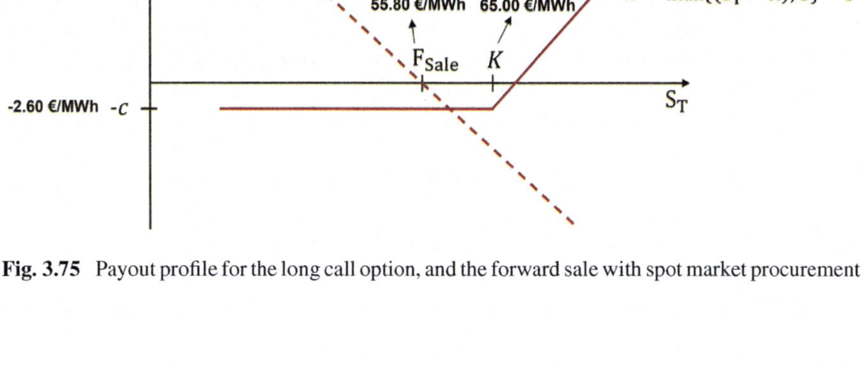

Fig. 3.75 Payout profile for the long call option, and the forward sale with spot market procurement

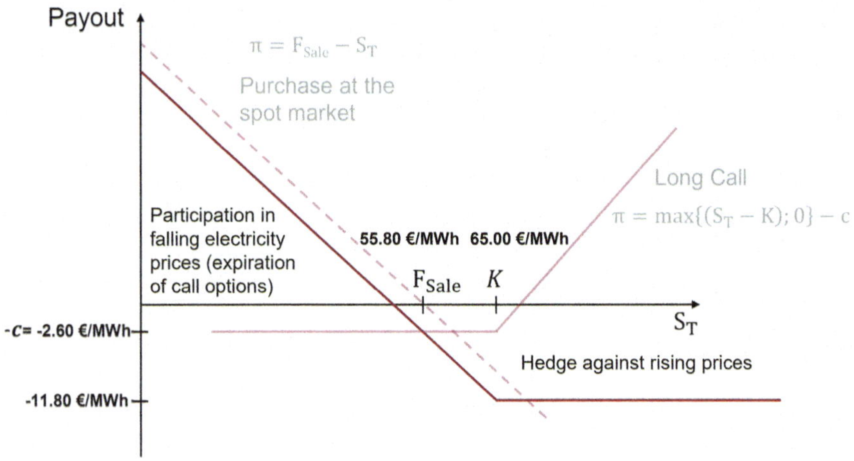

Fig. 3.76 Hedging with a call option

from the spot market procurement and the total payout profile is on a horizontal line, i.e. it does not fall any further. For every €/MWh the spot price increases, there is one €/MWh loss from the spot price procurement. However, there is also a gain of one €/MWh from exercising the option. This means that there is no longer a price risk for prices above 65 €/MWh. Nevertheless, the payout is negative because the option price had to be paid.

Unfortunately, these numbers are not unreasonable, and hedging with options is relatively expensive. Paying 2.60 €/MWh protects from the risk of a price above 65 €/MWh, which is an increase of nearly 10 €/MWh from the 55.80 €/MWh of the forward's contract sale. Obviously, the lower the strike price, the higher the option premium. Hedging against the risk of a price above 60 €/MWh would be considerably more expensive.

Strategy 4 is the so-called zero-cost collar. First, like in strategy 3, a long call is used to hedge against rising prices. When assuming the same parameters as in strategy 3, this can be done at a cost of 2.60 €/MWh as option premium. The idea of the zero-cost collar is to avoid the cost of the option premium by selling a put option that has the same option premium. This put has a profile like in Fig. 3.35, but with a premium of 2.60 €/MWh and a strike price of (in this case) 50 €/MWh. The premium is received for selling the option at the cost of losses when prices are low.

Figure 3.77 shows that payout profiles of the two options. The combined payout profile is equal to the addition of the blue curve (short put option) and the red curve (long call option). The resulting payout profile is shown in Fig. 3.78. If the spot price S_T lies between the two strike prices, the payout is zero. There is a positive payout for prices above 65 €/MWh due to the call option, but if the price falls below 50 €/MWh, we do not benefit due to the put option which will be "exercised against us".

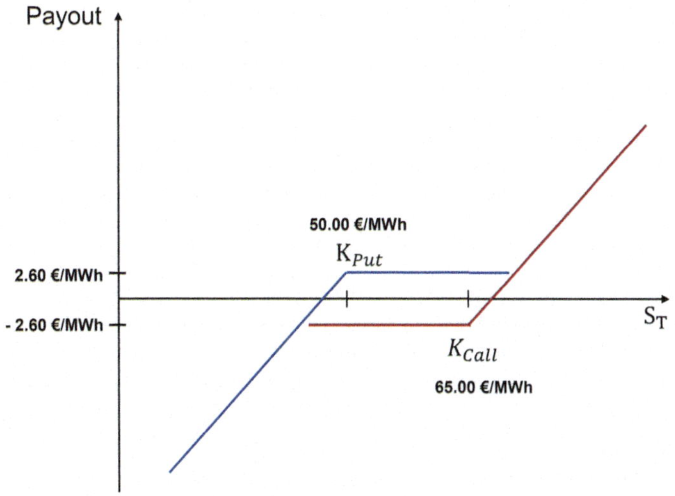

Fig. 3.77 Long call and short put payout profile

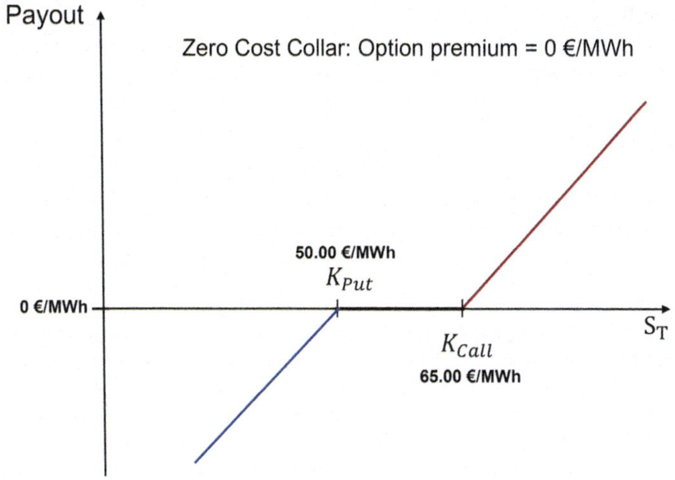

Fig. 3.78 Zero-cost collar payout profile

Table 3.16 gives three examples of the profit of this strategy, depending on the spot price S_T. In the first case, none of the options is exercised, and the profit is zero. In the second case, the price is above the strike price of the call option. Therefore, it is exercised. Since the price is above 50 €/MWh, the buyer of the put option will not exercise it, and the profit is 5 €/MWh. Finally, in the third case, the call option is not exercised, but the buyer of the put option will exercise it and sell something for 50 €/MWh that costs 45 €/MWh on the market. This means, there is a loss of 5 €/MWh.

Table 3.16 Zero-cost collar results

No.	Spot price S_T(€/MWh)	Strike price call (€/MWh)	Strike price put (€/MWh)	Profit/loss (€/MWh)
1	60	65	50	0
2	70	65	50	5
3	45	65	50	−5

In a next step, this payout profile for the zero-cost collar is added to the profile for the spot market procurement payout. Similar to strategy 2, this is because the two options are purely financial instruments and physical procurement on the spot market is required. This is shown on the left-hand side of Fig. 3.79. The sum of these two payout profiles gives the total payout profile of the zero-cost collar strategy, as shown on the right-hand side of Fig. 3.79.

If prices at maturity are lower than 50 €/MWh, the strategy leads to a profit of 5.80 €/MWh. This profit does not increase further with falling spot prices (and thus lower procurement costs on the spot market), because the put is exercised and every additional €/MWh that is earned from spot price procurement is used to compensate the buyer of the put option.

In the price range from 50 to 65 €/MWh, the payout decreases diagonally with a gradient of 45 degrees. Since in this price range neither the put nor the call is exercised and the two option premiums offset each other, the total profile is equal to the spot market procurement strategy profile. At 55.80 €/MWh the profit is zero, and if the price keeps increasing, this leads to losses. If the price rises to 65 €/MWh, there is a loss of 9.20 €/MWh. If the price rises above 65.00 €/MWh, the call option is exercised, which protects from losses due to any further price increase.

In conclusion, selling the put compensates the costs of buying the call. By doing that, potential gains in the low-price range are given up for protection from price risk in the high-price range. Unfortunately, in the practical setting, this is usually not symmetric. This mostly results from the bid-ask spread (see Sect 2.4.2) of options, which are often not very liquidly traded. Consequently, more upside potential must be

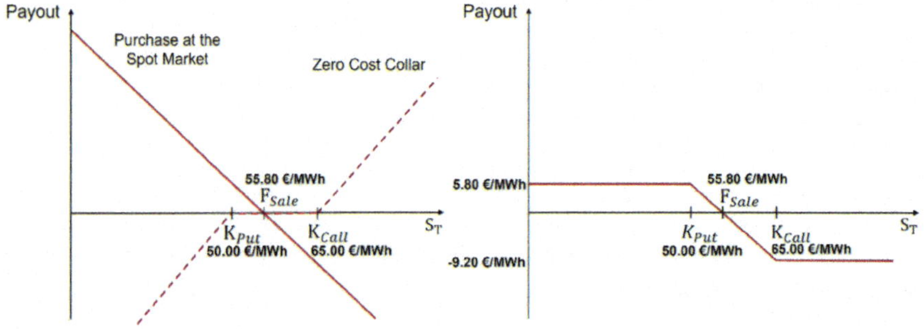

Fig. 3.79 Covering a sale with a zero-cost collar

traded away to compensate for the downside risk hedge and the costs for "insurance" (here: the cost for buying the call) are typically higher than gains made from selling opportunity (here: the profit from selling the put). Nevertheless, a zero-cost collar is an interesting product for hedging against price fluctuations while retaining a certain amount of downside and upside leeway. Furthermore, this strategy is used by some financial players when they enter a new market. They use the zero-cost collar to hedge against extreme market events ("black swan events") but actively trade and take risk within these limits.

3.4.1.2 Option Valuation

So far, we have always assumed that there is a certain, externally imposed, option price or option premium. In the following we will show how this price can be calculated. It is beneficial to understand how the option value is determined, especially because optionality is often inherent in certain energy-related contracts. For example, we have already showed the connection between swing options and flexible power plants. A comprehensive understanding of option valuation thus enables informed decision-making.

The price of an option reflects the value that the buyer of the option has by being able to exercise it but does not have to. This price, i.e. the value of optionality, can be derived fundamentally and can be divided into the following two components:

- **Intrinsic value**, which is the current market price of the underlying minus the strike price. When positive, it is the payout of the option if it were exercised immediately. The intrinsic value can be zero or positive.
- **Extrinsic value** (sometimes also called time value or volatility value), which is the value resulting from the asymmetry of the payout profile. Due to this asymmetry, the payout of an option can be very high, but it has limited downward risk.[81] This is because the option can go deep into the money, but the option payout cannot be negative, as it is not exercised when the option is out-of-the-money. Therefore, the extrinsic value is always strictly positive.

The option value is the sum of intrinsic and extrinsic value. Which of the two is higher depends on the circumstances: If the intrinsic value is already very high, the extrinsic value can be low and vice versa. For instance, if an option is out-of-the-money and thus the intrinsic value is zero, the extrinsic value can still be high and in this case amounts to one hundred per cent of the value of the option.[82] In all cases, the extrinsic value is zero on the day of maturity.

The value of an option depends on a number of factors:

---

[81] Look again at the payout profiles of a long call (Fig. 3.31) and a long put (Fig. 3.34).

[82] Nevertheless, it can be very close to zero, depending on the option parameters.

- **Price of the underlying**
  - A call option becomes more attractive with a price increase. Consequently, the option price increases with the price of the underlying.
  - A put option becomes less attractive with a price increase. Therefore, the option price decreases with the price of the underlying increasing.
- **Strike price**
  - The higher the strike price, the lower the probability that a call option will be in-the-money (and hence exercised) at maturity. Therefore, the price of a call option decreases with an increasing strike price.
  - For the same reason, the price of a put option increases with the strike price and vice versa.
- **Remaining term of the contract**
  - The value of an option (both call and put) increases with an increasing remaining contract duration because the probability of an increase of the intrinsic value then also increases. In other words, there is more time for the underlying price to move in the direction desired by the buyer. Since the payout profile is asymmetric, a higher contract duration increases the value of the option.
- **Volatility**
  - As the volatility of the price of the underlying asset increases, the option price rises because—analogous to the remaining term of the contract—the probability of an increase in the intrinsic value increases. Again, since the payout profile is asymmetric, a higher volatility increases the value of the option. This applies to both call and put options.

There are several methods to assess all these factors and determine an option value. One way is to use analytical models which have been developed for this purpose. One of the most established models is the "Black–Scholes" model. This model consists of analytical formulas to evaluate options. The following is an example of the Black–Scholes formulas. Equations (3.21) to (3.23) show how the price c of a European call option can be calculated depending on the price of the underlying S and the current time index t:

$$c(S,t) = SN(d_1) - Ke^{-r(T-t)}N(d_2) \tag{3.21}$$

with

$$d_1 = \frac{\ln(S/K) + \left(r + \frac{1}{2}\sigma^2\right)(T-t)}{\sigma\sqrt{T-t}} \tag{3.22}$$

$$d_2 = d_1 - \sigma\sqrt{T-t} \tag{3.23}$$

where

$N(\cdot)$	Distribution function of the standard normal distribution, i.e. $N(0, 1)$.
S	Current price of the underlying
t	Time index
T	Exercise time (European call)
r	Risk-free interest rate
K	Strike price
σ	Standard deviation of the price of the underlying.

This formula is analytically true if and only if the underlying's price movement follows a normal distribution, which is rarely the case in practice. Nevertheless, since it can be easily implemented in a spreadsheet or a pocket calculator, an analytically precise value of an option price can be determined if all the necessary parameters, in particular the standard deviation, are known.

Side Remark

The theory behind the Black–Scholes model was published in 1973 by F. Black, M.S. Scholes and R.C. Merton (the latter working somewhat independently). This theory quickly gained influence and Scholes and Merton were awarded the Nobel prize in economics in 1997. Sadly, Black passed away in 1995 and was not awarded the prize.

In 1998, Scholes and Merton failed spectacularly as directors of the hedge fund Long-Term Capital Management (LTCM) due to liquidity issues. LTCM's business model was to invest money applying financial principles involving spreads and optionality. Unfortunately, at some point in 1998, there was a very quick market movement, and they failed to switch positions fast enough.

This is an example of a market lacking liquidity, and the associated liquidity risks. In this case, in 1998, the market was very illiquid, and it became hard or virtually impossible to change positions. This led to the bankruptcy of the hedge fund. The event caused a financial crisis, and Alan Greenspan[83] lowered the US federal funds rate because of it.[84]

Apart from the Black–Scholes model, other analytical models have been developed. In addition to analytical models, there exist alternative approaches to evaluate option prices. As an example, we will discuss one numerical method, the Binomial model, in the following.

The binomial model uses a decision tree starting at $t = 0$ when the price of the underlying is P_0, as shown in Fig. 3.80.

The tree follows price development paths for the remaining time of the option, split in discrete periods. P_0 can increase by a factor u or decrease by a factor d at the end of each period (also called steps).[85] This is repeated for each period and Fig. 3.80 shows the result for two time periods: The resulting price from the previous period is multiplied by u for an increase or d for a decrease.

[83] Chair of the Federal Reserve of the United States at the time.

[84] Fleiss and Kumaar (2022).

[85] The binomial model allows exactly these two cases.

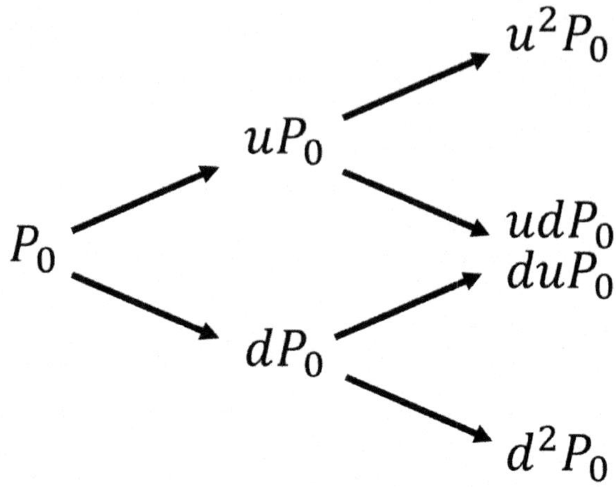

Fig. 3.80 Decision tree of the binomial model

Next, each price development scenario can be assigned a probability, and the fair option value is calculated as the probability-weighted average of the discounted values. If this model is calculated for 30 periods, this results in 1.07 billion possible price paths.

The extent of the price increase or decrease and the probabilities depend on the price volatility of the underlying, i.e. its standard deviation σ, and the period duration, according to the following equations[86]:

The factor of the price increase:

$$u = e^{\sigma\sqrt{\Delta t}}. \tag{3.24}$$

The factor of the price decrease:

$$d = e^{-\sigma\sqrt{\Delta t}} = 1/u. \tag{3.25}$$

The probability of increase:

$$p_{\text{increase}} = \frac{e^{r\Delta t} - d}{u - d}. \tag{3.26}$$

The probability of a decrease

$$p_{\text{decrease}} = 1 - p_{\text{increase}}. \tag{3.27}$$

[86] These parameters follow from the modelling assumptions and in particular the fact that log price changes are considered.

In the following example, we will show how these parameters can be used to calculate the price of an option with the binomial model. The market parameters of underlying and the call option are assumed to be the following:

- Future price in $t = 0$ (F_0): 20 €/MWh
- Strike price (K): 22 €/MWh
- Volatility (σ): 35%
- Interest rate (r): 5%
- Remaining term: 3 months
- Time of one period (Δt): 1 month (0.08333 years).

Calculation of the model parameters:

- $u = e^{\sigma\sqrt{\Delta t}} = e^{0.35\sqrt{0.08333}} = 1.1063$
- $d = 1/u = 0.9039$
- $p_{\text{increase}} = \frac{e^{r\Delta t} - d}{u - d} = \frac{e^{0.05 \cdot 0.08333} - 0.9039}{1.1063 - 0.9039} = 0.4954$
- $p_{\text{decrease}} = 1 - p_{\text{increase}} = 0.5046.$

The model parameters can now be used to create a tree of possible price developments, as shown in Fig. 3.81.

Equation (3.28) shows as an example of the calculation for the value in the blue field (two times up and one time down):

$$F_0 \cdot u^2 \cdot d = 20 \cdot 1.063^2 \cdot 0.9039 = 22.13. \tag{3.28}$$

In a next step, the possible prices of the underlying in Fig. 3.81 are converted to the option payouts. This means that the option payout formula is applied to each of those values: The payout of a call option is the price of the underlying minus the strike price of 22 €/MWh, if that is positive, or zero otherwise. This leads to the results shown in Fig. 3.82. These are the intrinsic values of the option.

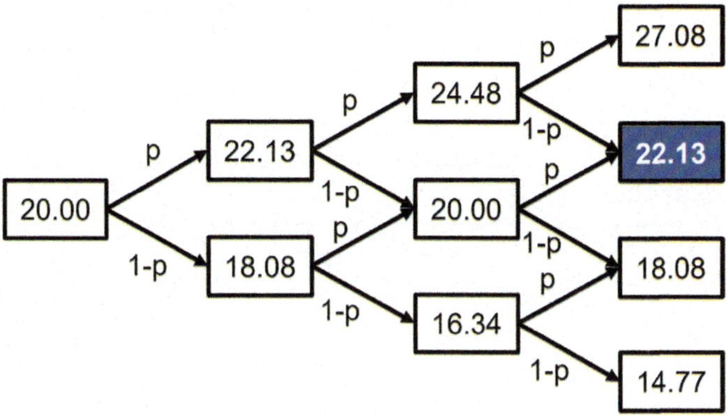

Fig. 3.81 Tree of possible price developments using the binomial model

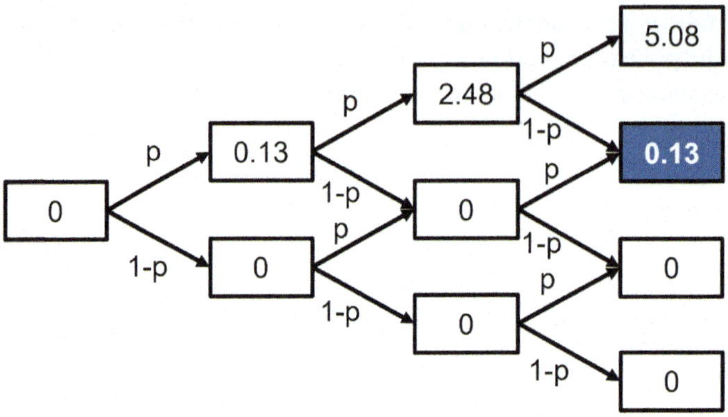

Fig. 3.82 Binomial tree of intrinsic values

In a final step, the fair value of the option is calculated. To do this, the tree is filled in reverse order from right to left. For that, all possible realisations in the rightmost column at maturity are taken as a starting point. Now the values in each "mother node" are calculated as the discounted probability-weighted sum of its two "successors" on its right. This is done starting with the rightmost column going to the left until the root node. The result is shown in Fig. 3.83.

For example, Eq. (3.29) shows how to calculate the value in the blue field in Fig. 3.83:

$$\begin{aligned}&[5.08 \cdot \mathrm{p} + 0.13 \cdot (1 - p)] \cdot e^{-r\Delta t} \\ &= [5.08 \cdot 0.4954 + 0.13 \cdot 0.5046] \cdot e^{-0.05 \cdot 0.08333} = 2.57. \end{aligned} \tag{3.29}$$

As a result, the root node of the binomial tree contains the fair option value of 0.66 €/MWh according to the binomial method.

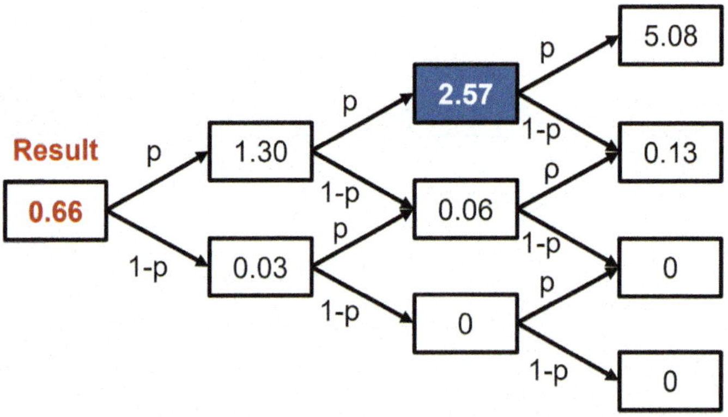

Fig. 3.83 Calculation of the fair option price using the binomial tree

Another numerical approach to determine option values is the Monte Carlo method. Monte Carlo originates from stochastics where random samples of a distribution are repeatedly drawn. Such simulation models require high modelling effort and high computational power. The methodology behind Monte Carlo simulation for calculation option values is beyond the scope of this book.

3.4.1.3 Separating Financial and Physical Risk with Swaps

An advantage of swaps as well as futures and options as financial products is that they separate price risk from physical risk: the risk of high prices in a supply situation can be separated from the risk of not getting physical delivery. Two different contract partners can be selected, one for the mitigation of price risk and another for physical trades.

Assume for instance a municipal utility that has bought a forward contract for natural gas. There is a physical risk inherent in this contract because it is possible that the supplier cannot deliver. If it is also a fixed price contract, price hedging is also not applicable in this case, because the fixed price already *is* the price hedge.

Alternatively, the municipal utility can hedge these two risks with a supply contract at a variable (market) price and additionally a swap. If the delivery fails, the price is still be secured because it is decoupled from the physical delivery contract. This is shown in Fig. 3.84 with the gas supply contract on the right-hand side and the swap to protect from increasing prices is on the left-hand side.

During a force majeure event, physical supply may be lost, but price protection remains in place. This means that if the gas supplier does not supply the gas, the municipal utility must look for another supplier and conclude a new contract. The gas price in the new contract will be based on current future prices for natural gas, which may have risen since the original supply contract was concluded (which is not unlikely in the case of force majeure). However, the municipal utility will continue to pay the fixed price agreed in the swap. When the two risks are combined in a fixed price supply contract, a force majeure event results in the loss of both physical supply and price protection. The municipal utility will have to look for a new supplier—which may be hard especially during a force majeure event.

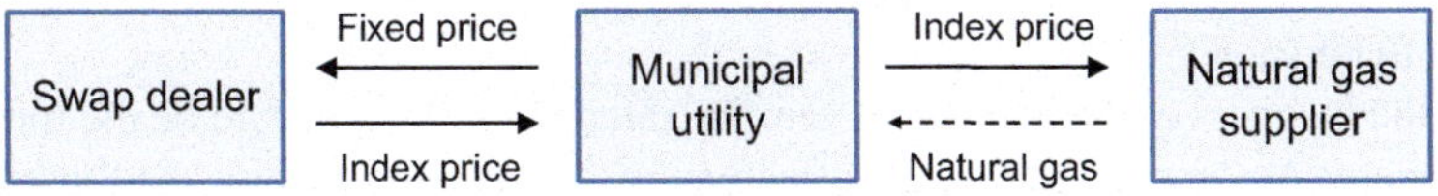

Fig. 3.84 Example of a swap for price risk protection of gas delivery

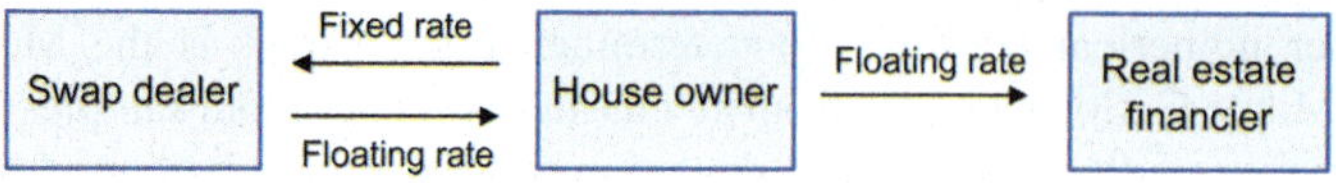

Fig. 3.85 Floating rate to fixed rate swap

3.4.1.4 Interest Rate Swaps

Swaps originally arose in the financial world in the context of interest payments. Typically, an interest rate swap is used to pay off some kind of debt someone has from a fixed rate to a floating rate or vice versa. Such swaps are also very common in energy markets.

This principle can be well explained using the example of a property sale. If someone buys a property as a private person and takes out a loan for it, this can be done at a fixed interest rate. This is then usually tied for a certain period (e.g. five or ten years). However, there is also the possibility of taking out loans that have a variable interest rate instead of a fixed rate. This variable interest rate is then based on the development of the base rate (e.g. it corresponds to the base rate plus x per cent). The borrower, however, runs the risk of suddenly having to pay more for the real estate loan if the base rate rises. He might have thought that was a good thing when the loan contract was signed, but over time his life situation may have changed in such a way that he would now prefer a fixed interest rate. However, loan agreements for house purchases cannot usually be changed retrospectively, and if they can, then only at a high cost. In such a case, he can conclude a swap like shown in Fig. 3.85.

In this case, the original loan (right-hand side of the figure) is converted from a floating rate to a fixed rate by using a floating-for-fixed swap. This means that in the swap the house owner receives a floating rate for paying a fixed rate (left-hand side of the figure). In sum, instead of a floating interest rate the net payment now is a fixed interest rate thanks to the swap: The house owner pays a fixed rate to the swap dealer, receives a floating rate in return and gives that floating rate to the original real estate financier. Of course, a financial volume must be specified in the swap, to reference the deal size and to calculate the cash settlement amounts. In this example this would be the debts to the real estate financier.

There is also the opposite, a fixed-for-floating rate swap, where instead of a fixed interest rate a variable interest rate is paid. To do this, a bank that will do such a swap needs to be found.

In the business context, a "plain vanilla" interest rate swap could be the following agreement: A company makes an agreement with a bank to receive the three-month EURIBOR[87] and pay a fixed rate of 5% per annum every three months for three years on a notional principal of €10 million. Schematically, such a swap between a company and a bank looks like in Fig. 3.86.

[87] EURIBOR stands for Euro Interbank Offered Rate. EURIBOR is the average interest rate at which many European banks lend each other euros.

Fig. 3.86 Plain vanilla fixed-for-floating interest rate swap

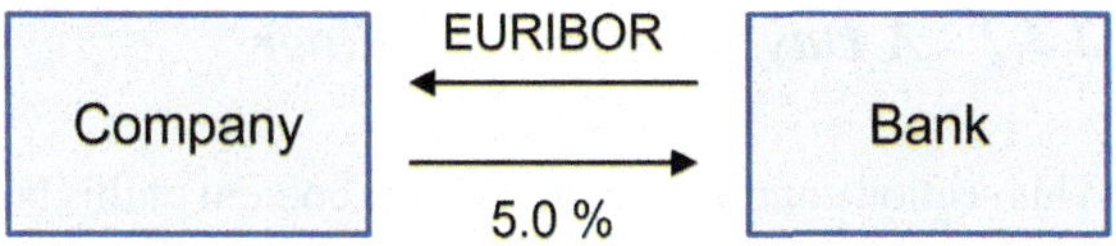

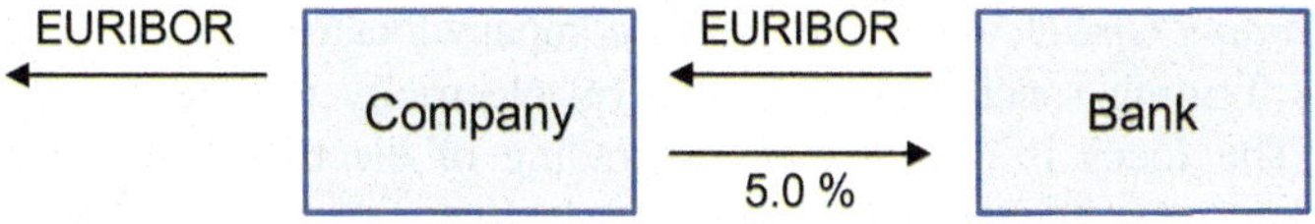

Fig. 3.87 Net fixed payment

The reason for entering such a swap from the company's point of view may be a new strategy. In this case, the payment of a fixed interest rate now seems better to the company than a variable interest rate such as EURIBOR. So if the company previously serviced a loan of €10 million with a variable interest rate (EURIBOR), with this swap it converts the loan into a fixed interest rate of 5%, as shown in Fig. 3.87.

In net terms, the company now pays a fixed interest rate of 5.0% instead of the variable EURIBOR rate. The variable interest payment of the original loan still exists but is converted into a fixed interest rate for the company through the swap.

As an alternative to financing, where a fixed interest rate is exchanged for a variable interest rate, such a swap can of course also be used to manage the payments from an investment. Take for instance a variable-rate investment, where the payments fluctuate and are based on an interest rate index such as EURIBOR. If someone wants to change the payment from this variable interest rate to a fixed interest rate without giving up the investment, an interest rate swap can be used. In this case a regular variable payment is traded for a regular fixed payment. Since the variable payment is received from the investment itself, the result is a fixed interest rate. The dates on which the swap payments are made should of course correspond to those of the investment.

A swap involves cash flows to be made regularly in the future. It can therefore be regarded as several combined futures. The payments do not take place only once, but once a month, once every 3 months or once a year. So, in addition to the formula for how the payment is calculated and the principal, a swap also contains a payment schedule, e.g. on the first of each month for three years. Of course, one could alternatively conclude a corresponding number of futures or forwards. However, this is much more time-consuming and often the corresponding markets are not liquid enough. In contrast, swaps only have to be negotiated and concluded once.

3.4.2 A Day on a Trading Floor

This section contrasts the theoretical content of this book with a hands-on description of the activities on a trading floor. The trading floor is where all the tasks discussed so far take place. We present a case study of a typical day on the trading floor of *Impuls Energy Trading GmbH*, an energy trading company located in Cologne, Germany.

Impuls Trading specialises in short-term electricity trading on the European markets. The focus is on the efficient trading of electricity—particularly from renewable energy sources (RES)—on the short-term energy markets, which are characterised by high volatility and constant price fluctuations.

Impuls Trading also utilises all the markets presented so far in this book (forwards/futures and spot/intraday markets). The spot and intraday markets are particularly important for the marketing of electricity from renewable energies. For this reason, Impuls Trading focuses on short-term marketing. The daily work is embedded in the daily marketing process.

The day on the trading floor usually begins **between 8:00 and 9:00** with the forecast preparation for the RES portfolio. This process is partially automated to relieve traders of repetitive operational tasks. The forecasts are continuously updated every quarter of an hour for the next few days and stored in an internal database. The trader checks whether the forecasts for the current and following day are plausible and processed correctly in the portfolio management system. To do that, the trader also takes a look at weather forecasts and the forecasts for the portfolio of all photovoltaic and wind systems in the areas with marketed PV or wind generation units.

After that, the resulting open positions of the entire portfolio are exported from the portfolio management system and entered as orders in the hourly auction of EPEX spot. This must be done **by 12:00 noon** at the latest. Figure 3.88 shows an example of a typical price-independent offer for marketing the output from a PV portfolio.

In the day-ahead auction, electricity can be bought or sold for the following day in hourly resolution (see Sect 2.2). For buy and sell orders, a price limit can be specified up to which the order is to be executed. In Fig. 3.88, unlimited orders were submitted: The orders are executed at any price between −500 EUR/MWh (the minimum price) and 4000 EUR/MWh (the maximum price) and thus independent of the price. The red numbers in Fig. 3.88 represent the order volume. In hour 08–09, for example, 5.9 MW are to be sold.

The bids can be submitted individually or as a so-called block bid. There are different forms of block bids. For example, base or peak load block contracts are typical in short-term trading. A block bid means that only the entire electricity generation of a power plant from morning to evening is sold or nothing at all. The bid would therefore only be accepted as a whole (i.e. all hourly bids) or not at all. However, block contracts play a rather minor role in the marketing of renewable energy plants.

Comment	Period (cet/cest)	-500,0	4.000,0
	00 - 01	0,0	0,0
	01 - 02	0,0	0,0
	02 - 03	0,0	0,0
	03 - 04	0,0	0,0
	04 - 05	0,0	0,0
	05 - 06	0,0	0,0
	06 - 07	0,0	0,0
	07 - 08	-1,4	-1,4
	08 - 09	-5,9	-5,9
	09 - 10	-11,2	-11,2
	10 - 11	-16,8	-16,8
	11 - 12	-21,4	-21,4
	12 - 13	-19,0	-19,0
	13 - 14	-16,4	-16,4
	14 - 15	-11,9	-11,9
	15 - 16	-5,4	-5,4
	16 - 17	-0,5	-0,5
	17 - 18	0,0	0,0
	18 - 19	0,0	0,0
	19 - 20	0,0	0,0
	20 - 21	0,0	0,0
	21 - 22	0,0	0,0
	22 - 23	0,0	0,0
	23 - 00	0,0	0,0
	Sum	-109,9	-109,9

Fig. 3.88 Example offer from a PV portfolio

After bid submission, the exchange processes all individual bids for all 24 hours of the following day in its auction and subsequently calculates the market prices for the hourly spot products. Figure 3.89 shows the price formation for a specific hour on a particular day in the German market area.

The upper part of the image shows the entire supply curve in green, and the entire demand curve in orange. In the lower part of the image, the intersection of the supply and demand curve is shown enlarged. In this case, the exchange has calculated a price of 73,52 €/MWh for a quantity of 33.935 MW. The exchange publishes the auction results **at around 12:40**. If the company's price offer was below this price,

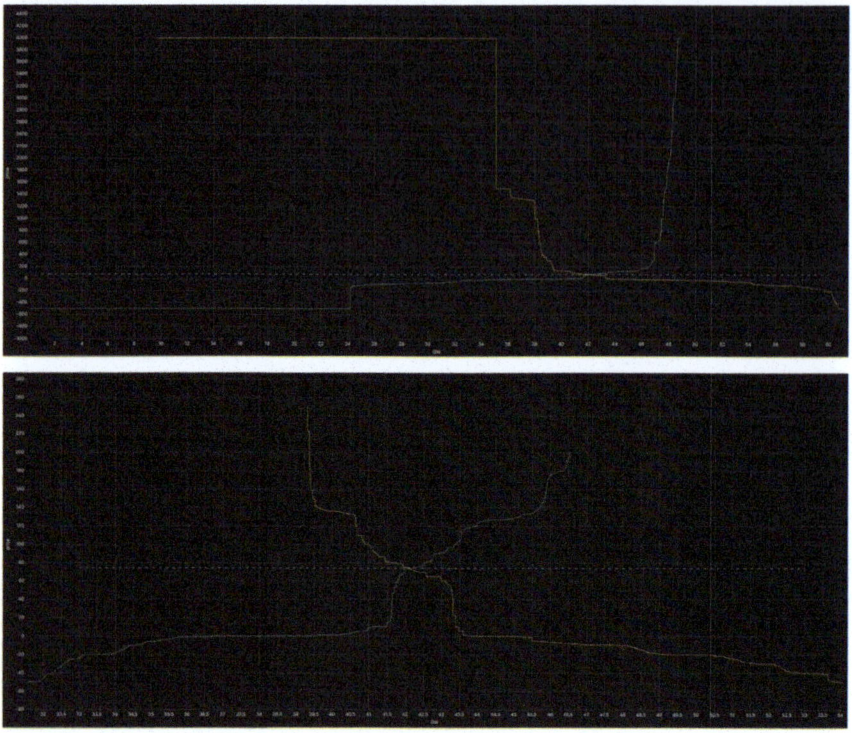

Fig. 3.89 Example supply and demand curves on the day-ahead market[88]

the capacity from the portfolio is sold. If it was higher, the plant's expected output has not yet been marketed.

Once results become available, the trading company starts a post-order process, in which the exchange results are imported into the company's portfolio management system. This process is typically automated, but the traders still have to check whether the results have been imported correctly and the data is plausible. If everything has worked successfully, the day-ahead position is balanced, which means that the quantities still to be marketed are displayed as zero in the system at hourly level. Once this process is complete, it is time for a lunch break.

The next deadline is **at 14:30**, when the next day's schedules must have been sent to the transmission system operators. Impuls Trading has generation plants in the regions of 4 transmission system operators. The schedules are sent automatedly and directly from the portfolio management system to the four TSOs in encrypted form. Since the schedules may change due to intraday trading activities and changing forecasts, they are updated and sent every 15 min. Continuous monitoring is particularly important here, as it involves the physical booking of energy between the individual balancing groups.

[88] Source: Own illustration of EPEX Spot (2023)—Impuls Trading GmbH.

Trading companies typically obtain both weather information and power forecasts from third parties. Some companies develop their own forecasting algorithms in addition. The forecasts are thus adapted to the requirements of the company's own portfolio. For example, it is particularly challenging to create forecasts for PV systems with a high proportion of self-consumption. Here, it is necessary to forecast the generation of the PV system on the one hand, but also the potential self-consumption on the other hand, to be able to market the best estimate for the grid feed-in. Such individual forecasts are currently not available in sufficient quality from the established forecasting companies and must therefore be prepared individually.

Once the schedules have been sent, the traders prepare for the intraday auction that takes place **at 15:00**. In the German market bidding system, all generation and consumption volumes must be marketed and procured on a quarter-hourly basis. The hourly marketing in the 12 o'clock day-ahead auction results in residual quarter-hourly quantities that can be traded on the market for the first time in the quarter-hourly intraday auction. Together with the new positions arising due to updated weather forecasts, these volumes are placed in the intraday auction. This auction runs almost exactly like the day-ahead auction, with the only difference that now only 15-min products are traded.

After the intraday auction, as with the day-ahead auction, the exchange results are imported into the company's portfolio management system and the updated schedules for the next day are sent to the TSOs. Once the processes have been successfully completed, the portfolio management system again displays an open position of zero.

Continuous intraday trading then begins **at 15:00** for the hourly contracts and **at 16:00** for all quarter-hours of the following day. The lead times in intraday trading have been continuously shortened over time and are now only 5 min in Germany. This means that positions can be traded up to 5 min before the start of delivery. For example, if a participant wants to buy 20 MWh of electricity for the quarter of an hour from 16:00 to 16:15 and finds a seller who is willing to sell this quantity, the offer must be accepted by 15:55 at the latest. This only applies if both parties are in the same control area, while trading across control areas can be carried out up to 15 min before delivery. This is the last time to transfer energy between the control areas by means of a so-called external schedule.[89] The main reason for such short lead times is the increasing feed-in of fluctuating energy sources, which makes long-term feed-in forecasts more difficult. It helps electricity traders to trade their electricity volumes at shorter notice and simplifies portfolio management. On the other hand, it requires fast reaction times from traders and flexible portfolio management processes.

For the trading floor, this means that from now on all traders will permanently monitor the weather forecast for the next day. Various information is used for this purpose, such as satellite images, wind and solar forecasts from various providers, but also public information from grid operators about expected demand and feed-in to the grid and available cross-border capacities to other countries.

Due to the constantly changing weather forecasts, but also for other reasons such as unforeseen outages of individual plants in the portfolio, the open position for all

[89] This is the current status in November 2023.

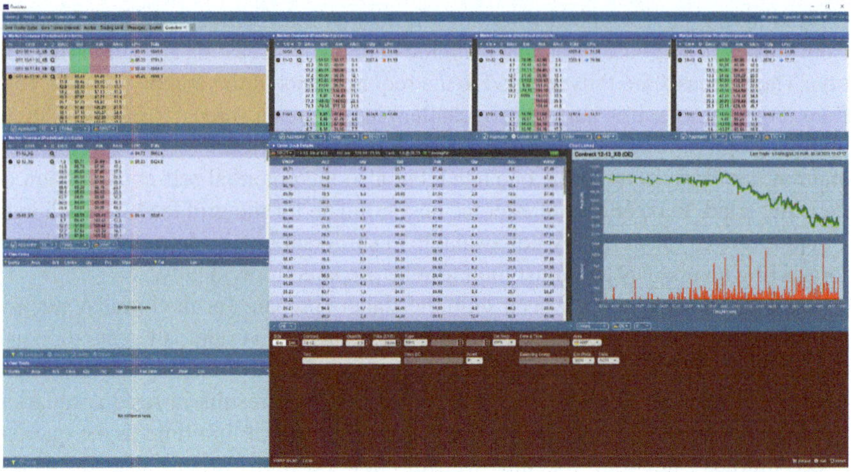

Fig. 3.90 Example of a ComTrader screen

quarter-hours is constantly changing. The main aim of intraday trading is to minimise shortfalls or surpluses in the balancing group through short-term trading activities to meet the forecasting obligations of the balancing group contract and reduce balancing energy risks. In view of increasingly flexible plants such as e.g. battery storage, short-term trading can also be used to produce electricity from plants at short notice in line with demand, which brings both economic and system-stabilising benefits. Although wind and photovoltaic systems are supply-dependent generators, these systems are often shut down at low (i.e. negative) market prices. In this way, even inflexible generation plants such as wind and PV ensure price-based mitigation of extreme prices in times of strong electricity oversupply.

Impuls Trading uses the ComTrader access of EPEX Spot for this purpose. Figure 3.90 shows a trader screen of the ComTrader.

In contrast to day-ahead trading, prices are set on the intraday market using the "pay-as-bid" method. This means that if seller and buyer agree on a certain price, the trade is performed at that price. If, for example, a trader receives information that an open position has arisen, for instance due to a changing weather forecast, he attempts to close it again via the intraday market. To do this, he looks at the supply and demand situation for the corresponding quarter of an hour, which is reflected in the respective order book.

In continuous trading, all buy and sell orders are collected in an order book.[90] A trade is executed when orders on the bid side and the ask side offer the same (or better) price. A trader has the option of using limit or market orders. A market order is unlimited and is executed at any price. In other words, placing a market order

[90] This is the same principle as with long-term contracts, which was explained in Sect 2.4. The main difference is that the delivery period, i.e. the product, is now no longer the base or peak contract for a specific period in the distant future, but a 15-min time interval in the near future.

means that one is willing to sell at the lowest possible or buy at the highest possible price. A limit order is only executed if an order from the other market side offers a price that is equal to or better than the limit order price.

Traditionally, such trades are executed by humans and not computer algorithms. This has several advantages. Human traders are able to draw on intuition and experience to interpret complex market conditions and make informed decisions. They can adapt quickly to changing market situations and react flexibly in critical situations, avoiding risky trading decisions that algorithmic systems may not recognise. Especially in certain niche markets or exotic trading strategies, human expertise can be invaluable as it requires in-depth specialist knowledge and individual adaptability.

However, algorithms are increasingly being used, particularly for trading decisions that do not require a high degree of intelligence or where execution is the main focus. Impuls Energy Trading, for example, uses a position closer algorithm to automatically close portfolio deviations based on forecast adjustments.

Impuls Energy Trading does not currently market any flexibility on the balancing power/energy market. Nevertheless, the balancing power/energy market is important. As the last market-based instance for the short-term balancing of supply and demand, the transmission system operators procure flexibility via the balancing power and balancing energy market (see Sect. 3.3.4.2). The cost of activating this flexibility is passed on to those responsible for the imbalances via the imbalance energy price.[91] The balancing energy market provides important information for traders on the intraday market as a shortage signal of supply and demand. Moreover, since marketers of renewable energy plants tend to be strongly correlated with the imbalances and systematically have to bear balancing energy costs, the potential costs of balancing energy (which are based on the balancing energy price) are an indicator of the financial risk of deviations in their own balancing group.

Figure 3.91 shows an example of the bidding curve for the aFRR capacity auction for a specific time block. On the left are the bid prices for negative balancing capacity, on the right for positive balancing capacity. In the lower part of the image, the centre part is shown enlarged.

The image shows that both the price curve for negative and positive balancing power are very flat in the centre. This means that many suppliers offer similar prices and thus "guess" the correct market price. This also makes sense, as pay-as-bid applies here, i.e. everyone receives their bid price and not the uniform market price as in the day-ahead auction. Traders with sufficient experience can obtain important information about the current market situation from these curves and their changes over time and integrate this information into their bidding strategies. This shows once again that all relevant markets must be considered in an integrated manner in order to successfully market flexibility.

Even energy traders usually do not spend the whole evening in their office. However, as soon as the last trader has called it a day, trading continues. Wind

[91] This process differs from country to country, but usually the balancing costs are part of the imbalance price.

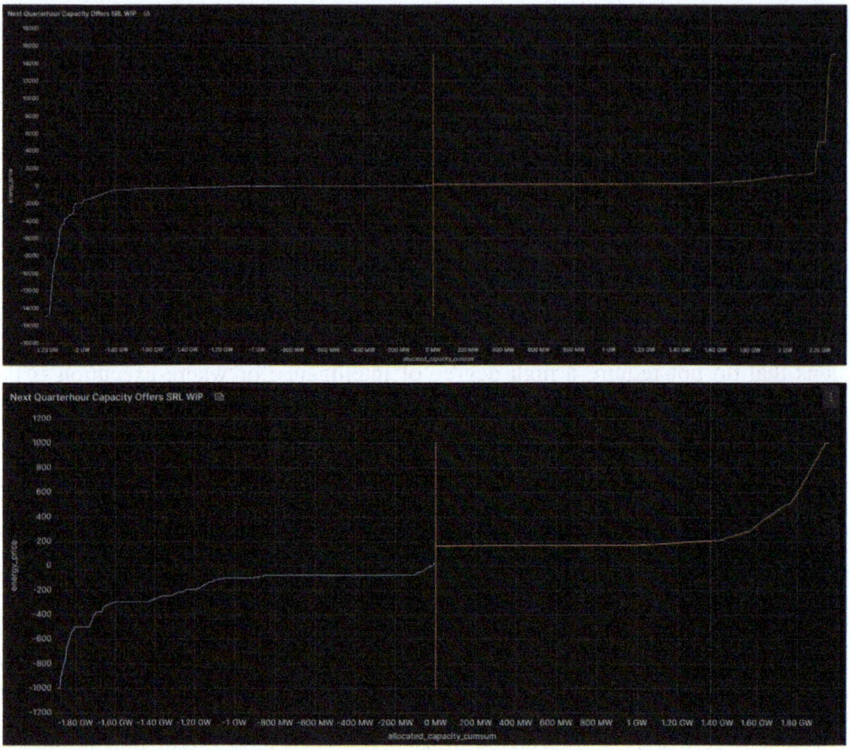

Fig. 3.91 Example supply and demand curves on the balancing power market[92]

energy trading in particular makes it necessary to operate a 24-h trading floor. Intelligent performance forecasts in conjunction with good trading algorithms now make it possible to implement 24/7 trading relatively cost-effectively—with the positive side effect that traders only have to be on standby and no longer have to work night shifts. At Impuls Energy Trading, therefore, only the traditional, human-manned trading closes **at 19:00**, but computer-based trading activities continue automatically around the clock. Overall, the processes in the short-term can be described as a continuous passing on of the baton. It is important that all available information is integrated and processed in the best possible way in order to provide both traditional traders and algorithms with the best possible basis for decision-making.

[92] Source: Own illustration of Impuls Trading GmbH based on data from Netztransparenz.de (2023).

3.5 Performance Measurement

The more complex a company's activities become and the more sophisticated its trading activities in different markets and with different products are, the more difficult it becomes to measure success or failure. The key variable to measure success in energy trading is the value of the portfolio—including the value of the open position. Section 3.5.1 introduces the underlying concept.

Furthermore, companies are interested in attributing departments 'or even individuals' success in the process. We refer to this process as performance measurement within the organisation (Sect. 3.5.2) and suggest to use wholesale market prices as internal transfer prices within the organisation.

3.5.1 Performance Measurement for the Whole Company

Assume a company has carried out several trading activities in a sample year. Since January, all sorts of different deals were made, and sometime around July the company wants to know how well they are doing, i.e. what is the current value of the trading portfolio.

The value of the trading portfolio has two components: The first is realised revenues, i.e. revenues from deals where positions have been opened and closed. The second is the value of the open position at market prices on the day the performance measurement takes place. The value of the open position thus answers the question how much the company would need to pay and receive as payments if the open positions would be closed on that particular moment. We will calculate both components in the following and introduce the concept of the hourly price forward curve in the process, which is an essential tool to evaluate the open position.

3.5.1.1 Value of Closed Deals

The first value component of a portfolio is simply the revenue from all deals which have been signed by the company. Every contract bought and sold is added up. The amounts the company receives for a commodity sold have a positive sign (as they lead to revenues), and amounts for anything the company has bought have a negative sign (as they lead to costs).

In our example, the starting point thus is to review all the deals that have been done during the last six months. By looking at everything that was sold and bought (including volumes and prices), profits and losses during this time can be calculated.

However, an important revenue component needs to be added, namely the value of the open position.

3.5.1.2 Value of the Open Position

Why does the value of the open position have to be added to the value of closed deals? This can be demonstrated using a simple example: Assume that the company has only sold energy up to July, but has not yet bought anything. Summing up all contract values for closed deals would result in a huge revenue. However, this says nothing about whether the company has been successful. The company has amassed a large short position—which has to be closed before the delivery period begins. This will lead to payments, which need to be included in the analysis.

Therefore, the current value of the open position at market prices needs to be calculated. This is done via mark-to-market (M2M). The open position's value depends on the future contract prices for the delivery period. The total value of the open position is the potential income from the sale of long positions minus the potential costs for closing short positions. Since electricity is balanced hourly, the open position often includes both long and short positions. During certain hours a company may be long and in others, they may be short.

To calculate the value of a portfolio, it is easy to include energy already procured and sold. This can simply be done by adding sales revenues and subtracting purchase costs. However, the current value of the open position also needs to be computed. This is done using Eq. (3.30).

$$v = \sum_{h} OP_h \cdot p_h \tag{3.30}$$

The value of the open position $\boldsymbol{v}$ is equal to the sum over all hourly intervals (or 15 min intervals) of the open position. The open position in each hour, OP_h, is measured in energy amounts, i.e. megawatt hours, and it is multiplied by the forward price for that hour. Typically, the open position varies from hour to hour and even from quarter-hour to quarter-hour. For the sake of simplicity, we will assume an hourly open position.

The differences in the open position for each hour result mostly from the fact that purchases are done with standard products, e.g. base or peak, whereas the sales follow customer-specific hourly consumption profiles. Therefore, the difference between purchases and sales also varies from hour to hour.

Equation (3.30) shows that for each hour the open position (in terms of energy amounts) is multiplied by the price of that hour. It is possible to calculate—at least approximately—the open position for each hour of the delivery period. However, there is usually no individual price for every hour of next year. This makes it somewhat challenging to come up with a precise value for the open position, obviously.

However, we have already learned that prices for liquidly traded products, in particular standard products such as base and peak, are readily available. An example for this is the price for the front-year baseload contract. That is the price for 1 MWh delivered in each of the 8760 hours of next year. As the open position varies from hour to hour, hourly prices are needed, and these cannot simply be taken from the front-year baseload contract alone. The solution to this problem is an Hourly Price Forward Curve (HPFC).[93]

3.5.1.3 Hourly Price Forward Curves

Since hourly products are not traded on the wholesale futures market and there are thus no publicly available hourly prices, an HPFC needs to be derived from available prices (such as the baseload future price) and other available information (such as the monthly, weekly and daily price structure; the general principle how to do that will be explained below). Having such an HPFC, we can go back to Eq. (3.30) and calculate the value of the open position for each hour.

Before calculating an HPFC, it is important to determine what it will be used for. There are many different possible applications for HPFCs, such as

- Calculation of the current value of a trading portfolio.
- Pricing of electricity supply contracts.
- Evaluation of load profiles.
- Evaluation of power plants.
- Making (dis)investment decisions.
- Derivation of trading strategies.
- Risk management.
- …

Depending on the application of the HPFC, there are different approaches to the calculation. For example, if the HPFC is used to determine the value of a power plant with many utilisation hours, price peaks that last only one or a few hours are not particularly important. It is much more important to determine periods when the power plant is in-the-money and when it is out-of-the-money, and how much money it can earn on average during these hours. In contrast, the value of a flexible storage device such as a battery or a flywheel depends on the price of electricity in each individual hour or even quarter of an hour. Therefore, the focus of such an HPFC should be on hourly price volatility rather than the average price of several consecutive hours.

Furthermore, the method of calculating an HPFC depends on the desired time horizon. An HPFC that covers only the next few days is completely different from an HPFC that covers a few months or even years. This results from the different information available. The longer one looks into the future, the less information is available that can be used to calculate the HPFC. This information can for instance be

- Load structure (seasonal, monthly, weekly, daily and hourly),
- Power plant park (including renewables),
- Hourly feed-in from supply-dependent energy sources (e.g. wind and solar),

[93] Usually, on trading floors people just use the abbreviation.

- Fuel prices,
- etc.

An example where the time horizon is particularly relevant is the hourly feed-in from wind power: there are good estimates for hourly wind feed-in for the next few days, but not for the next few weeks or months. In other words, forecast errors increase with time to delivery (see e.g. (Nguyen and Müsgens 2022)). Since this feed-in usually has a large impact on the hourly electricity price, an HPFC that only covers the next few days can use this information, but an HPFC for the next year cannot.

All these aspects need to be considered when an HPFC is developed. In the following, we will present a general approach how to calculate an HPFC. For reasons of standardisation, we will always refer to an hourly HPFC and not a quarter-hourly (q)HPFC. In principle, all concepts mentioned also apply to quarter-hourly price forward curves. While HPFCs can be bought from an external provider, understanding the underlying concept is valuable.

Regardless of their application, all HPFCs share some common characteristics and are similar in design. There are basically two necessary properties that each HPFC must fulfil:

(1) Arbitrage freedom: The average price of the HPFC's hourly prices corresponds to all respective future prices observed on the market.
(2) Shape: The shape of the hourly prices reflects their expected structure.

From these basic properties, one can derive other specifics of an HPFC, e.g. that it must include

- a seasonal profile,
- a weekly profile,
- continuous price developments,
- independency of historical extreme events,
- etc.

The starting point for the calculation of the HPFC are current market prices. These are typically available on the websites of exchanges, such as e.g. EEX or Nord Pool. They provide prices for the different calendar years' standard products in the future, including daily or even more frequent price updates for these annual contracts.

As the base price is the average price of the 8760 hours of the delivery year, the 8760 hourly prices of the final HPFC must (see property (1)) on average be equal to the future base price for the respective year. The same holds for the peak hours[94]: the average price of all peak hours in that year must equal the future peak price for that year.

[94] See Sect 2.1.2.

The exchanges also provide prices for quarters, months, weeks and even weekends. For calculating the HPFC, one can (and should, see property (1)) also use the additional information contained in these prices. If a future price for the forecast period is available and it is not used, information is lost, and hence, the quality of the HPFC decreases. If, however, quarterly future prices are used, the quality increases. This is because for one year instead of having only one base price there are now four base prices. These are usually different for each quarter as a result of seasonality. The volume-weighted average of the four quarter prices in a year will be equal to the price of the whole year at a given reference date. In other words, if quarterly prices are available, they should always be used, because they contain more information and thanks to the seasonality will make the HPFC look more realistic (see property (2)). However, quarterly prices are typically only available for the next few years and their liquidity decreases further into the future. Hence, if the HPFC should cover several years, yearly future prices are also needed.

Additionally, there are monthly price quotations, which should also be included in the calculation of the HPFC when available (see again property (2)). However, the available monthly prices will cover an even shorter period than quarterly prices. Liquid trading for monthly products typically just happens for the next few months. In the middle of the year, probably even January of the front year will not be very liquid, but price quotations may still be available.

Based on this information, an "integrated" forward curve, as shown in Fig. 3.92, can be created with the peak prices in red and the base prices in blue. This curve is derived from the information previously mentioned, i.e. yearly, quarterly, and monthly prices. At the short end of the curve, i.e. in the near future, there is a monthly shape due to the availability of monthly price quotes. When there are no more monthly price quotes available, quarterly price quotes are used and at the long end of the curve, i.e. far ahead, the shape of the curve is based only on yearly price quotes. This is how the market values the next years today.

If other price quotes are available, these should also be used to guarantee arbitrage freedom and an expected shape. Such price quotes can be weekly or weekend prices. This is however somewhat challenging from a modelling perspective for the following reason: Arbitrage freedom must be guaranteed for all future prices. This is easy to deal with for monthly, quarterly, and yearly prices, because they will always be arbitrage free if they are taken from an exchange or another (trustworthy) trading platform like a broker. This will be illustrated with an example. Take for instance the price quotes in Table 3.17.

The volume-weighted average price of the four quarters[95] equals the price for the whole calendar year. This calculation is straightforward because quarters and years do not overlap. However—and this is why it is challenging to include weekly and weekend prices—weeks, weekends, and months often overlap. If weekly (and

[95] The year is a leap year, hence it contains 8784 hours. The number of hours of each quarter is simply the number of days multiplied by 24. However, due to the time change, the first and the last quarter have one hour less and one hour more respectively.

Fig. 3.92 Sample forward curve[96]

Table 3.17 Price quotes for sample calendar year and quarter years (base prices)

Period	Price in €/MWh
Cal-y	133.29
Q1/y	137.65
Q2/y	121.07
Q3/y	129.06
Q4/y	145.28

Table 3.18 Exemplary price quotes for sample months and weeks (base prices)

Period	Price in €/MWh
August	76.19
September	84.07
Week 31	66.54
Week 32	70.75
Week 33	76.34
Week 34	82.26
Week 35	83.75

weekend) prices shall be included in the HPFC, additional effort is required to guarantee arbitrage freedom. This is illustrated with the price quotes in Table 3.18.

The price quotes in the table cover two calendar months (August and September) and five calendar weeks. In this example, weeks 31 to 34 lie in August. Week 35 lies

[96] Source: Own illustration based on data from EEX.

Table 3.19 Base and peak price quotes September 2023

Period	Price in €/MWh
Base September	93.27
Peak September	101.00

partly in August (28th to 31st) and partly in September (1st to 3rd). For calculating a HPFC that starts in August and includes the price quotes of the weeks 31 to 35, the HPFC must satisfy all the following arbitrage freedom conditions:

- The average price of all hours within a week must equal the respective weekly prices.
- The average price of all hours within August must equal the August price.
- The average price of all hours within September must equal the September price.

Since week 35 overlaps with both August and September, this poses a challenge because changing the hourly prices, e.g. for September 1st, affects both the average price of September and the average price of week 35. Imagine now that the HPFC's September and weekly prices are all arbitrage free, but the average price for August is a little higher than the August price quote of 76.19 €/MWh. This can be handled by lowering hourly prices at the beginning of week 35 (which lies in August) and increasing hourly prices at the end of week 35 (which lies in September) to make August arbitrage free and keep week 35 arbitrage free. However, this will also increase the average price of September and now the September prices aren't arbitrage free anymore. This becomes even more complex if also peak and weekend prices (base and peak) are considered.

Once a monthly/quarterly/yearly price forward curve as in Fig. 3.92 has been obtained for base and peak, the information can be combined into one curve. If an hour belongs to the peak, it receives the peak value. If an hour does not belong to the peak period, the off-peak price is relevant. The off-peak price can be obtained by considering arbitrage freedom conditions and using the prices for base and peak hours of the respective period. This is illustrated in the following example (Table 3.19).

The off-peak price for September is obtained by using formula (3.31). It states that the volume-weighted average of peak and off-peak prices within a period (in this case September) must equal the base price.

$$\text{hours}_{\text{peak}} * p_{\text{peak}} + \text{hours}_{\text{offpeak}} * p_{\text{offpeak}} = \text{hours}_{\text{base}} * p_{\text{base}}. \tag{3.31}$$

Since September has 720 base and 168 peak hours in the example, solving it for p_{offpeak} leads to

$$p_{\text{offpeak}} = 720^{*}93.27\,€/\text{MWh} - 168^{*}101.00\,€/\text{MWh} = 90.92\,€/\text{MWh}.$$

If peak and off-peak prices are used and applied to Fig. 3.92, the result is the price structure shown in Fig. 3.93.

Fig. 3.93 Price structure information of HPFC

The figure shows the hourly prices of the first four months (September to December) of a price forward curve and already provides a first somewhat reasonable approximation of hourly future prices. Seasonality is clearly present in the profile and prices in peak hours are higher than in off-peak hours.

However, up to this point, only information on peak and off-peak hours have been used. The next step adds additional information, in particular a more refined hourly structure. This information can come from different sources, e.g.

- Historical hourly price developments over the past years.
- Publicly available fundamental knowledge about the future, that will shape hourly prices (phase-out of certain technologies, changing household user behaviour due to increasing electromobility, changing industrial user behaviour because of switch to hydrogen, etc.).
- Assessments of professionals.
- Other sources.

The straightforward approach to develop an hourly shaping of the HPFC is to take historical hourly price information. While future hourly prices will (most likely) not be on the same *level* as historical prices, they are likely to have a similar *shape*. Hence, we can now utilise historic price data in hourly resolution—but must maintain arbitrage freedom when using the information. The first step is often to convert monthly average data for peak and off-peak into weekly data, i.e. to transform the forward curve in Fig. 3.93 to a weekly structure.[97] As a result, there are now weekly future prices (peak and off-peak) for the HPFC.

[97] A very straightforward approach can simply be the assessment that all weeks within a certain month have the same average price. Prices in weeks that lie partly in one month and partly in the next must be adjusted in order to guarantee arbitrage freedom.

Table 3.20 Weighting matrix of daily prices for an HPFC

Mon	Tue	Wed	Thu	Fri	Sat	Sun
108%	110%	111%	110%	109%	81%	71%

Then, in a next step, one can determine the relative price level for each day of the week[98] within a week and then the relative price level of a certain hour within a day, based on historical data. The former could, e.g. look as presented in Table 3.20.

The percentage values in the table correspond to the price of a certain day within a week relatively to the average price of the whole week. For instance, the average price level of all hours on a Monday are 108% of the weekly price.

Next, an hourly structure is needed. This can also be derived from historical data, such as the hourly price development of the single hours of a day over the last couple of years. Table 3.21 gives an example of such a historical price structure.

The table shows the possible result of an hourly price structure derived from historical hourly prices. In this case, the percentage values correspond to the price in the respective hour in relation to the daily base price.[99] For example, the price in hour 6 of a Thursday corresponds to 85% of the daily base price. Once such a matrix is derived from historical data, it can simply be applied to daily prices and the result will be a HPFC.[100]

Obviously, this methodology applies some simplifications. First, it assumes that the shape of hourly prices in the future will correspond to their shape in the past. This neglects additional information one might have, such as fundamental knowledge about future changes influencing the price structure (changes in capacity, etc.). If, for example, a large additional amount of PV is expected, it is reasonable to assume that prices during daytimes will be significantly lower than they currently are. The same applies to other generation or consumption technologies that might shape the hourly price curve. Second, particularly in the short term, there is more information on expected hourly prices available than contained in historical prices. For example, if overcast skies are forecasted for the next few days, the daytime price dip will probably be smaller than in the past. Nevertheless, assuming that the general "behaviour" of hourly prices is similar to their behaviour in the recent past leads to a good and robust estimator of an HPFC. The whole process of deriving an HPFC can thus be summarized as presented in Fig. 3.94.

Once the HPFC has been calculated, it can be used to calculate the value of any open position using Eq. (3.30). If the open position (in megawatt hours) in every

[98] This is because the hourly price development within a week usually follows the same pattern every week: Prices are higher on weekdays and tend to be lower on weekends, when demand is lower.

[99] For optimal results, these percentage values should be in relation to the peak and off-peak price, because these have already been developed at this point. However, for illustrative purposes an hourly structure is shown relative to the base price, because the typical hourly shape of a day is more easy to see.

[100] In practice, probably some further adjustments would be needed to guarantee arbitrage freedom.

Table 3.21 Weighting matrix of hourly prices for an HPFC

Hour	Mon (%)	Tue (%)	Wed (%)	Thu (%)	Fri (%)	Sat (%)	Sun (%)
1	76	82	85	87	88	107	108
2	70	78	80	82	82	99	98
3	68	75	77	78	79	95	93
4	66	73	75	77	78	91	88
5	67	75	78	78	80	90	87
6	78	82	84	85	87	91	86
7	103	104	104	105	107	95	86
8	122	121	122	121	124	104	89
9	127	126	126	125	128	109	93
10	117	115	115	114	117	107	91
11	107	105	105	105	107	98	85
12	100	100	99	99	101	92	83
13	94	93	93	94	94	84	75
14	91	91	90	90	88	75	62
15	90	91	90	90	86	73	58
16	95	95	94	94	90	79	68
17	102	100	100	100	96	91	85
18	119	116	114	113	111	111	116
19	130	126	124	122	120	126	142
20	137	132	130	128	123	131	153
21	127	123	120	119	116	125	149
22	114	109	107	107	106	115	141
23	107	101	99	100	102	112	139
24	93	88	87	87	91	100	124

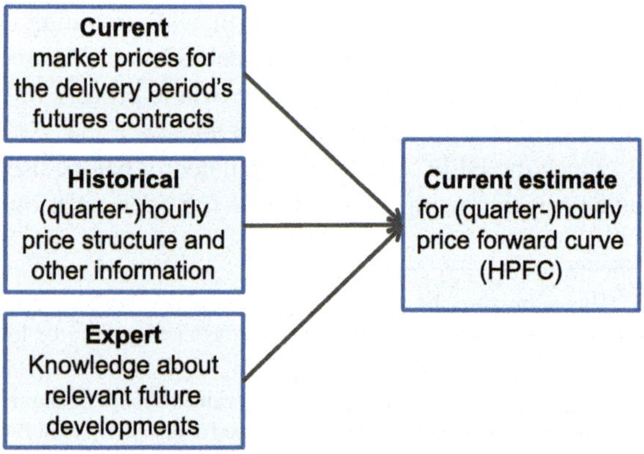

Fig. 3.94 Process for formation of an HPFC

hour is known, this can be multiplied with the price of that hour and the sum of all these hourly values is the current value of the open position.

Note that the HPFC is also a valuable tool to price any schedule, e.g. a consumer specific load profile. Furthermore, it can be used as input into the power plant optimisation described in Sect. 3.3.5.

3.5.2 Performance Measurement Within the Organisation

Utilities are also interested in determining who within the organisation contributed how much to the company's success. This can be valuable information for two reasons: First, companies want to identify the strengths and weaknesses within their organisation to manage resources optimally. In particular, it seems natural to learn from past mistakes and take corrective measures in under-performing parts of the organisation. Second, portfolio management is a high risk and high reward area, where employees are often incentivised with performance-based remuneration. Hence, the company needs a quantitative measurement of performance within the organisation.

For an integrated portfolio, it may be hard to determine where the money is coming from, i.e. whether it comes from the power plant, the trading team, or the sales team with its retail customers. We demonstrate the challenge using the example of an energy utility company with a retail focus first. In this case, the company requires a differentiation of success between procurement division (responsible for procuring the energy on the wholesale market) and the sales division (responsible to sell the energy to final consumers).[101]

Now assume in this example, we are interested in how well a sales department did when they closed a deal with a retail customer. Assume the following:

- The sales department sells a base product to a retail consumer for 55 €/MWh.
- The procurement department purchased the corresponding position for 50 €/MWh.
- The profit contribution is 5 €/MWh.

The profit contribution for the company is 5 €/MWh, which is easy to derive from the two contracts. The question however is, whose success drove the profit. Is the sales department getting a bonus or should the procurement team handling the purchase get it? Uncertainty and perceived unfairness in such situations can cause conflicts within the company, and it is therefore important to have a transparent environment where trust and cooperation between colleagues are maintained. The overarching objective must always be to have a profitable company, a profitable department comes second.

Coming back to the question, the answer is mark-to-market or M2M. M2M means that a position is compared with its market value. The important information missing

[101] As we have mentioned in Sect. 3.2.4, the required skill sets are so divers for these two tasks that they are rarely performed by the same people (or even organisational unit).

at this point is the wholesale market price at the time of the sale to the retail customer because this is the reference price for both departments.

If we assume that the wholesale price at the time of the retail sale was 52 €/MWh, then selling the electricity for 55 €/MWh to the retail customer above wholesale market price and the difference is quantifiable. In a simplified way, we could say that the sales team was able to sell something for 55 €/MWh, which was worth 52 €/MWh on the market. Therefore, they did a good job. Their share of the total profit contribution of 5 €/MWh is thus 3 €/MWh.

In the example, the procurement team made an additional profit by buying something for 50 €/MWh that had a value of 52 €/MWh. Of course, this purchase cannot have been back-to-back, because if it were, it would have been purchased on the wholesale market, and the cost would have been 52 €/MWh. The procurement team's success came either by buying in advance when prices were low, or waiting after the energy was sold for the price to go down.[102]

At this point, the total profit contribution has been separated by departments. 3 €/MWh can be attributed to sales, 2 €/MWh to procurement.

Mark-to-market generally enables a clear ex-post division of monetary success and thus enables performance differentiation between departments. Rightly implemented, it can be broken down to the level of individual employees. This is often done with prop traders responsible for their own trading book.

In a second example, we separate the profit contribution between a generation department, managing a power plant, and a sales department, again selling to final consumers.

- The sales department sells a base product to an end consumer for 55 €/MWh.
- Generation produces at 23 €/MWh.
- The profit contribution is 32 €/MWh.
- The settlement price for the delivery period is 52 €/MWh.

The question is again: Is the profit a success of the sales department or of generation? We use the same methodology as before: mark-to-market. If the market price at the time of the sale is 52 €/MWh, then the profit contribution of the sales department is 3 €/MWh (55 €/MWh minus 52 €/MWh), while the generation has a value of 29 €/MWh (52 €/MWh minus 23 €/MWh).

On a broader scale, this methodology can help to identify the strengths and weaknesses of the organisation. However, we advise to do so with care. People within the organisation should be colleagues not competitors.

If mark-to-market performance measurement is not performed properly, this can lead to absurd results. For instance, there is an example of a large integrated company where it was more rewarding for the traders to negotiate internal transfer prices with the power plant division than "beating" the market. Thanks to very low transfer prices (well below market prices), the traders were able to make absurd bonuses

[102] Another possibility is that the purchase team found someone who was willing to sell for 50 €/MWh even though the market price was 52 €/MWh, but that is rather unlikely.

during some periods, not because they were trading geniuses but because they got cheap electricity from their internal counterparties.

Furthermore, meaningful marking-to-market is more complex than the above examples show. For instance, determining the current market prices of standard products seems simple, but has some interesting aspects: Does one use bid, ask or mid-quote? At what time exactly is the deal executed? When did the sales department sign the contract? When did the information reach the procurement team? Daily settlement prices? Depending on the specific situation of the company, there are some degrees of freedom to the optimal answer and hence there is not necessarily a one-size-fits-all solution. Furthermore, the value of an individualised schedule is often difficult to determine. This becomes even more complex if it contains some flexible elements. On top of that, the costs of equipment and labour, etc. need to be determined and properly assigned and profit centres should cover their respective share of overhead costs. Nevertheless, the mark-to-market approach should always be the basis of any performance measurement.

References

BMWi. (2015). *An electricity market for Germany's energy transition*. https://www.bmwk.de/Redaktion/Migration/DE/Downloads/Publikationen/weissbuch-englisch.pdf?__blob=publicationFile&v=2

Bundesnetzagentur. (2023). *Smard*. Accessed on December 27, 2023. https://www.smard.de/home

Elia Group. (2022). *Balancing services: mFRR—Design note on mFRR balancing service*.

ENTSO-E. (2018). *Electricity balancing in Europe–Guideline*. https://docstore.entsoe.eu/Documents/Network%20codes%20documents/NC%20EB/entso-e_balancing_in%20_europe_report_Nov2018_web.pdf

ENTSO-E. (2019). Annex 2: Alternative configurations of the Bidding zone review region "Nordics" which are to be considered in the bidding zone review process. *Bidding Zone Review Region Nordic Region*.

ENTSO-E. (2022). *Overview of frequency control in the Nordic power system*. https://www.epressi.com/media/userfiles/107305/1648196866/overview-of-frequency-control-in-the-nordic-power-system.pdf

ENTSO-E. (2023). *Frequency containment reserves (FCR)*. Accessed on December 1, 2022. https://www.entsoe.eu/network_codes/eb/fcr/

ENTSO-E. (2023). *PICASSO*. Accessed on December 1, 2022. https://www.entsoe.eu/network_codes/eb/picasso/

ENTSO-E. (2023). *TERRE*. Accessed on December 1, 2022. https://www.entsoe.eu/network_codes/eb/terre/

ENTSO-E. (2023). *MARI*. Accessed on March 1, 2023. https://www.entsoe.eu/network_codes/eb/mari/

EnWB. (2023). *Interconnector*. Accessed on December 1, 2022. https://www.interconnector.de/wissen/uebertragungsnetzbetreiber/

EPEX Spot. (2022). *Epex spot trading brochure*. Accessed on December 27, 2023. https://www.epexspot.com/de/node/117

EPEX Spot. (2023). Retrieved from https://www.epexspot.com/en by Impuls Trading GmbH.

European Commission. (2017). *Guideline on electricity balancing*. Accessed on December 27, 2023. https://eur-lex.europa.eu/legal-content/EN/TXT/HTML/?uri=CELEX:32017R2195

Fleiss, A., & Kumaar, A. (2022). *Why was long term capital management (LTCM) saved by the Government and lehman brothers left to fail?*. https://doi.org/10.2139/ssrn.4045798

Grothe, O., Müller, C., & Müsgens, F. (2006). *Modellierung von Energiepreisrisiken durch Bindefristen bei öffentlichen Ausschreibungen.* Zeitschrift für Energie, Markt und Wettbewerb, 59–62.

Müsgens, F., Ockenfels, A., & Peek, M. (2014). *Economics and design of balancing power markets in Germany.* International Journal of Electrical Power and Energy Systems, 392–401. https://doi.org/10.1016/j.ijepes.2013.09.020

Müsgens, F., & Steinhausen, B. (2010). *Portfoliomanagement: Optimale Energiebeschaffung unter Berücksichtigung von Risiken.* Zeitschrift Für Energiewirtschaft, 34, 109–116. https://doi.org/10.1007/s12398-010-0014-0

NASDAQ. (2023). *NASDAQ glossary.* Accessed on December 27, 2023. https://www.nasdaq.com/glossary/p/proprietary-trading

Netztransparenz.de. (2023). *Netztransparenz.de.* Accessed on January 4, 2024. https://www.netztransparenz.de/de-de/

Nguyen, T. N., & Müsgens, F. (2022). *What drives the accuracy of PV output forecasts?* Applied Energy, 119603.

Regelleistung.net. (2022). *Modalitäten für Regelreserveanbieter.* Accessed on January 8, 2022. https://www.regelleistung.net/xspproxy/api/StaticFiles/Regelleistung/Marktinformationen/Modalit%C3%A4ten_/Modalit%C3%A4ten_f%C3%BCr_Regelreserveanbieter_MfRRA/1__MfRRA_Lesefassung_Zielmarktdesign_(mit_Vergabefrist).pdf

Regelleistung.net. (2023). *Automatic frequency restoration reserve.* Accessed on July 1, 2023. https://www.regelleistung.net/en-us/General-info/Types-of-control-reserve/automatic-Frequency-Restoration-Reserve

Regelleistung.net. (2023). *Frequency containment reserve.* Accessed on December 27, 2023. https://www.regelleistung.net/en-us/

Regelleistung.net. (2023). *Manual frequency restoration reserve.* Accessed on July 1, 2023. https://www.regelleistung.net/en-us/General-info/Types-of-control-reserve/manual-Frequency-Restoration-Reserve

Regelleistung.net. (2023). *Regelleistung.net.* Accessed on December 27, 2023. https://www.regelleistung.net/en-us/

Scholz, D., & Müsgens, F. (2017). *How to improve standard load profiles: Updating, regionalization and smart meter data* (Bd. 2017 14th International Conference on the European energy market (EEM)). https://doi.org/10.1109/EEM.2017.7981939

Chapter 4
Risk Management

In this chapter, we first discuss the different types of risks associated with energy trading, as well as their characteristics and drivers. We then focus on price risk, credit risk and product liquidity risk management. After that, the concept of risk indicators will be presented. We end this chapter by explaining risk management processes.

Before we start a more detailed discussion, it needs to be stated that any economic activity implies risks. This is neither "good" nor "bad" but just the nature of business. The future is always uncertain. Hence, it should not be the objective of a company (or its risk management) to avoid all risks—as this would imply not to spend any money and result in the company's failure. Instead, a company needs to find the right balance between consciously taking risks (which implies chances) and keeping the company financially stable. Risks thus need to be taken purposefully, understood properly and managed professionally.

Risk management supports this process. In general, it focuses on the negative outcome resulting from risks manifesting. While this is a necessary endeavour, it should be clear that the company hopes for a more positive outcome at the same time. Otherwise, the company should refrain from an activity in the first place. Risk management in that sense is the "prepare for the worst" in the saying: hope for the best and prepare for the worst.

4.1 Types of Risks

The different types of risks, which are relevant for portfolio management in the energy sector, can be classified depending on their probability of occurrence and the extent of damage they can cause. Figure 4.1 shows a so-called risk matrix for the most relevant risks in this context.[1]

[1] Depending on a company's strategy, its situation, its exposures and its assets, the assessment of individual risks may vary.

F. Müsgens and A. Bade, *Energy Trading and Risk Management*,
https://doi.org/10.1007/978-3-031-57238-8_4

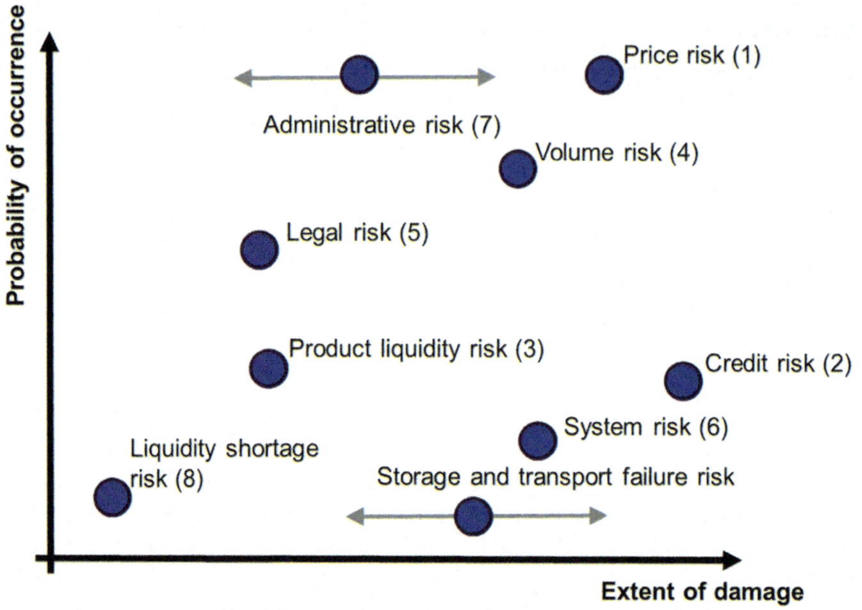

Fig. 4.1 Relevant risks in the energy sector

Of these risks, we have already mentioned two, namely price risk and credit risk. The need to have introduced them early on already confirms their status as important risks in energy trading. The figure visualises this with a combination of high probability of occurrence and high extent of damage for price risk. Credit risk has the highest extent of potential damage of all risks considered.

In the following, we will describe each of the eight[2] risks shown in Fig. 4.1 in more detail:

1. The value of an open position (both long and short) changes with price variations and can lead to a loss. The only exception is a completely balanced portfolio, i.e., the open position is zero. **Price risk** thus subsumes the risks of future price variations changing the value of the open position. Note that the value of the open position also changes the value of the whole portfolio (see Sect. 3.5.1), directly affecting the company's profit and loss function.

   Price risk is most influenced by two factors: the size of the open position and the volatility of forward prices. Price risk and the size of the open position are positively correlated because a larger open position (measured in MWh) causes a larger loss (or profit) for any given change in the forward price. The second factor influencing price risk is the variation of forward prices, in particular the variation's probability and intensity. Both probability and intensity can be described with price volatility which is typically estimated using statistical methods.

[2] We will not discuss storage and transport failure risk, because these are very specific for companies operating storage infrastructure.

The standard deviation is a common measure to estimate volatility. For instance, the volatility of n daily forward prices p_i can be calculated as the standard deviation of the daily (log[3]) returns as shown in the following equation:

$$\textbf{Daily volatility} = \sqrt{\frac{\sum_{i=1}^{n}(x_i - \overline{x})^2}{n-1}} \tag{4.1}$$

With x_i being either the daily returns $x_i = \left(\frac{p_i - p_{i-1}}{p_{i-1}}\right)$ or the daily log returns $x_i = \ln\left(\frac{p_i}{p_{i-1}}\right)$ on day i and $\overline{x}$ being the average of the daily (log) returns.

Another measure for price variation is the maximum expected or maximum occurred price change.

Price risk is probably the most common, most discussed and by most practitioners also considered the most important fundamental risk in energy trading. The management of price risk will be analysed in Sect. 4.2.

2. **Credit risk** measures the monetary loss that results from a counterparty default. It manifests whenever a company owes another company money but is unable to pay due to bankruptcy. Its likelihood is influenced by the credit rating of the counterparty. Such a rating essentially measures the probability of a counterparty going bankrupt (also referred to as default risk), and it can hence be used to decide with whom to trade how much.

   We have already mentioned that one advantage of futures trading is that the power exchange as counterparty has a very low default risk. With forward trading, on the other hand, the transaction takes place directly between two energy trading companies. This involves a low, but not negligible, risk of default.

   In addition to assessing and monitoring a counterparty's default risk, credit risk can be influenced by trading with several counterparties and limiting financial volumes for each of them. Don't put all your eggs in one basket—as the saying goes. With a diversified portfolio of counterparties, there is still a loss with the default of one of them, but it typically does not trigger a chain reaction that causes the bankruptcy of the creditor.

   There was a trading company operating in Belgium and the Netherlands that had a big client who went bankrupt and could not pay the bills. The loss of the trading company was so big that its parent company decided not to back that up and it had to go bankrupt. Even though the company was generally perceived as financially healthy, this one-time shock was beyond their ability to pay.

   Netting agreements can also help to reduce credit risk. If a company repeatedly trades with the same counterparty, a netting agreement specifies that all sales and purchases are netted, and the two companies owe each other the difference

[3] In the context of measuring volatility, using log returns often simplifies calculations and provides a more stable representation of the underlying financial processes. Log returns have other advantages such as being additive over time. This means that when you sum the log returns over different periods, you are effectively calculating the overall return for the entire period. Simple returns, on the other hand, are not additive.

only. If a counterparty defaults and there is no netting agreement, the insolvency administrator usually has the right to fulfil all contracts of value. This means that the (net) creditor only receives partial payment for the outstanding debts but has to pay all positions which are valuable to the insolvent debtor. In contrast, with a netting agreement the creditor only has to take care of the difference.

Finally, credit risk can be limited by using margin calls and collaterals. Having a collateral is like the simple example of getting a loan to buy a house: The bank requires some collateral, and in case the debt cannot be paid anymore, the collateral (i.e. the house) goes to the bank. A margin call is a demand to deposit additional funds or securities to cover potential losses.

The management of credit risk will be analysed in Sect. 4.3.

3. **Product liquidity risk** is the risk that the market for a product becomes illiquid. We have talked about liquidity in Sect. 2.6, and we mentioned that liquidity can be measured. Typically, a low bid-ask spread is a sign of high liquidity, another is the order book depth. Consequently, high product liquidity risk means that trading volumes become insufficient. Bid-ask spreads widen, the order book depth narrows and the whole market dries up. As a result, positions cannot be closed or opened without huge discounts.

   We will discuss the management of this risk in Sect. 4.4.
4. **Volume risk** refers to a deviation between expected and realised supply and/or demand volumes. One example is a utility planning to deliver a certain amount of energy to a client but being unable to do so due to a power plant outage. The electricity has been sold, but the power plant cannot produce it. Another example for volume risk is a change in the household customer base. As mentioned in Sect. 3.2, household customers have the right to change their supplier on short notice. Consequently, some may stay with their supplier while others may leave. A supplier can thus not know for certain how much electricity will be sold to household customers next year.

   Volume risk can also be influenced by consumer behaviour and depends on the type of supply contract. For instance, an open contract with an industrial customer where the delivery of any amount (or at least within certain thresholds) of energy needed is specified, the amount of energy supplied may differ from what was expected and procured. This is the volume risk for these customers.

   Additionally, forecast errors in demand and generation induce volume risk. The demand can deviate from the forecast due to a weather change. PV and wind plants also have stochastic behaviour. This is not the same as generation outages, because even if the equipment works well, the expected amount of electricity may not be generated due to changing weather conditions (more or less wind or sunshine than expected).
5. **Legal risk** is the risk that a contract or group of contracts does not contain the desired legal content. It occurs due to poorly written contracts, which lead to unforeseeable or non-fulfillable claims from third parties or obligations towards them. A company may have signed a contract believing that it contains a specific

deal that was found beneficial. However, the counterparty in the contract interprets the contract differently. As they are complying to their understanding of the contract, the outcome is not as desired by the first company—or vice versa.

In such situations, sometimes an arbitration case may be necessary. This is because in the energy sector there are often long-term contracts, some of them valid for up to 30 years. For instance, if one party invests in a power plant and a second party participates financially with a signed contract, changes in the market environment or changes in the regulation can lead to legal disputes about the contract.

For example, financial power plant contracts historically gave the right to nominate the day-ahead production schedules. In simple terms, such a contract sold a virtual slice of a power plant and the buyer was allowed to nominate it with a clean spark spread option. Assume now that markets have developed after the contract was signed and balancing power has become an important revenue source—while at the time the contract was signed, nobody talked about balancing power. As a result, the buyer of the slice may believe that he has the right to get additional revenues from balancing power through their virtual slices, arguing that they wanted to buy an approximation of the physical power plant and that the change in the market environment was unforeseeable. Thus, they believe to have the right to receive revenues from balancing power sales of the plant. On the other hand, the power plant operator who sold the financial contract could say that this is not in the written agreement. They sold specifically what is in the contract, i.e. the nomination of the production on the day-ahead market, and that they will fulfil the contract to the letter. Such situations require arbitration and getting the clauses of the contract interpreted by a court.

The influencing factors of legal risk are the design of the contract, regulatory changes affecting the markets, and the interpretation of the courts in arbitration cases.

6. **System risk** is the risk of malfunctions within the processes, systems and procedures of the company.

   For example, there may be problems with a company's software, or the hardware malfunctions or the company suffers from an electricity supply interruption preventing e.g. balancing the portfolio on the day-ahead market. As we have mentioned, buy and sell orders must be submitted to the power exchange until noon. These orders are often needed to balance the balancing group, which must be reported to the TSO soon after the auction. Imagine there is a power failure on the trading floor around noon and the schedule has not yet been submitted. This is a realistic scenario, and companies must be prepared for it. Some companies have, for instance, battery-powered notebooks or emergency power generators for these kinds of situations.

   This risk is influenced by the organisation and decisions related to the hardware and software to be used, e.g. the desired level of hardware redundancy or whether to use self-made software or professional software.

7. **Administrative risk** is the risk of incorrect information, making incorrect decisions, or performing incorrect activities due to human errors or non-compliance

with processes and rules. In other words, employees make mistakes in their professional activities.

Of course, mistakes will happen wherever work is being done. They are part of human nature, and unavoidable to a certain degree. For instance, we try to avoid spelling and grammar mistakes while writing this book. We use spell-check programmes and professional proofreading services. But still, it is not guaranteed that there are no mistakes anywhere in the book.

The effort required to avoid or minimise mistakes depends on how vital the specific sector is. An example of this is the aviation industry, where very strict procedures are applied to minimise human error.

Influencing factors on this risk are the training level of the employees, their motivation, the organisational procedures for documentation and communication, and the controls and double-checks in-place.

8. **Liquidity shortage risk (or cash flow risk)** is the risk that liquidity will not suffice to meet payment obligations. This may be due to poorly coordinated payment flows and deadlines, resulting in insufficient cash and cash equivalents.

   This is the case if a company needs to make a payment but doesn't currently have the money to do so. Compared to market liquidity risk, where the market needs liquidity, in the case of cash flow risk, the company runs out of liquidity. The business may be profitable overall, and its portfolio value may be positive. It may thus receive payments in the future, but that doesn't help right now. For instance, the payment from a forward contract that was sold is expected next year, but because of other payment obligations money is needed now. Due to poorly coordinated cash flows, the company cannot fulfil its payment obligation and has to declare bankruptcy.

   This risk can be minimised by drafting contracts with appropriate payment dates. Also, contracts may contain clauses that specify cash collateral in certain situations. Furthermore, the contract composition of the portfolio should be managed; i.e., a company can decide what kind of trading it does and what products and payment systems they use. It is also possible to ask a bank for a specific line of credit in advance to help cover immediate cash needs. If a business case is solid and the company profitable in general, a bank is likely to agree to a credit line.

4.2 Price Risk Management

The ultimate objective of price risk management is to avoid bankruptcy. Since future market developments are always uncertain, this requires methodologies and processes to deal with unfavourable developments. The most important one is the calculation and reservation of risk capital. The calculation of risk capital should result in a number the company could stand to lose in the worst case without threatening its survival. Price risk management monitors that the remaining risk capital (original risk capital minus realised losses) can cover any potential future loss. Doing so, the company can limit the potential damage by closing all open positions. While this would realise the loss, the loss is below the assigned risk capital making sure the losses stay below pre-allocated maximal values and the company survives.

This requires an estimate of the potential future loss that can reasonably be assumed. In that sense, price risk management is always forward-looking: make sure a potential future loss stays below the remaining risk capital.

Managing risk is thus never about total numbers, but about probabilities. It is necessary to determine both the financial risk of a business activity and the probability that the risk will materialise. This leads to the concept of risk indicators. A risk indicator is a measurable and observable factor or variable that provides insights into the likelihood or potential impact of a specific risk. These indicators serve as early warning signs, allowing us to take proactive measures to mitigate or respond to emerging risks. Risk indicators need to be calculated and monitored constantly. This will be shown in the following.[4]

As a starting point consider again the concept of risk capital: As we have mentioned, when engaging in risky trading businesses, risk capital is required, and this capital specifies how much a company can afford to lose. Once a company's risk capital is gone, it can close all open positions, realise the loss, and stop further trading. In practice, however, companies usually do not wait until the current portfolio loss exceeds risk capital before acting. Instead, other measures are taken before such an event happens. The question is, why is that the case and how do companies decide when to take measures and what measures that could be.

Simply said, if a company waits until losses are fully reflected in the portfolio value, it is not guaranteed that the loss stays below the assigned risk capital. This is because closing all open positions (which in this case happens when risk capital is already exceeded) can cause additional losses. In most cases, a company cannot act immediately. For example, the company needs time to come to terms with the fact that the losses exceed the risk capital. During this time, the losses can increase even further because of real-time price variations. Since the value of the open position depends on the forward prices, any change in the forwards or futures contracts can increase the losses.

This can be illustrated by looking at risk management in medium-sized trading companies: They typically have a daily risk management process implemented, i.e. risk parameters such as the portfolio loss are calculated using market settlement data. The calculation is often performed automatically overnight.[5] If the settlement price at the end of a day has reached a level where the risk capital is exceeded, the information is communicated internally (e.g. with a risk report) on the next morning. Before the position can be closed, the market may be completely different from the previous day.[6] Furthermore, when the trader starts to engage in the market, the

[4] In this section this is done based on the example of price risk management. Note that risk indicators may also be used in the context of other risks.

[5] This is often referred to as the *end-of-day process*. It usually starts when all day-to-day business has been completed.

[6] It is not unusual that prices in the morning are different from where they were the previous evening.

first price offers (particularly of illiquid products) can appear to be unreasonable and therefore he starts asking for price quotes via voice brokering, causing another delay. Time goes by and in the end, the position may be closed, but at a level that can be significantly above the risk capital. Particularly in extreme market situations (which often lead to limits being exceeded), losses can increase significantly until the open position is flat.

In conclusion, risk management needs to be proactive, and must look at what could happen to the open position one day (or even more days) in the future. This is the reason why companies do not wait until the risk capital is exceeded. Instead, they use risk indicators to assess the riskiness of their current open position. The most important risk indicators in the context of energy trading are presented in the following.

4.2.1 Stress Value

The stress value is the change in the value of a portfolio from one day to the next in an expected worst case. The stress value should not exceed the remaining risk capital. If the losses have not yet exceeded the risk capital today, but the stress value indicates that this could be the case tomorrow, there is a need for action.

Calculating the stress value requires an estimation of the worst-case daily price change. This can be approximated by taking historical prices and taking the maximum price change between two historical days. This price change is then applied to the current open position, and the result is the stress value. This is illustrated in the following example.

- (Assumed) worst-case price change: 15%.
- Open position: 10 MW base for calendar year y (leap year with 8784 hours).
- Current market price: 29.14 €/MWh.
- Stress: 29.14 €/MWh $\times$ 10 MW $\times$ 8784 h $\times$ 15% = €383,949.

In this example, the stress value, i.e. the worst-case loss from today to tomorrow, is €383,949. In other words: If the worst-case price change assumption of 15% is correct, the company's daily loss will not be higher than this value. If, for example, the company's current loss is €700,000 and the risk capital is €1 million, the open position must be proactively closed, as the total loss would exceed the risk capital if this worst-case price change were to occur.

4.2.2 Value-at-Risk

The Value-at-Risk (VaR) is the maximal possible loss a portfolio can suffer under normal market conditions, with a given probability (e.g. a confidence level of 99%) and within a specified period (e.g. one day). The VaR determines the loss of a trading

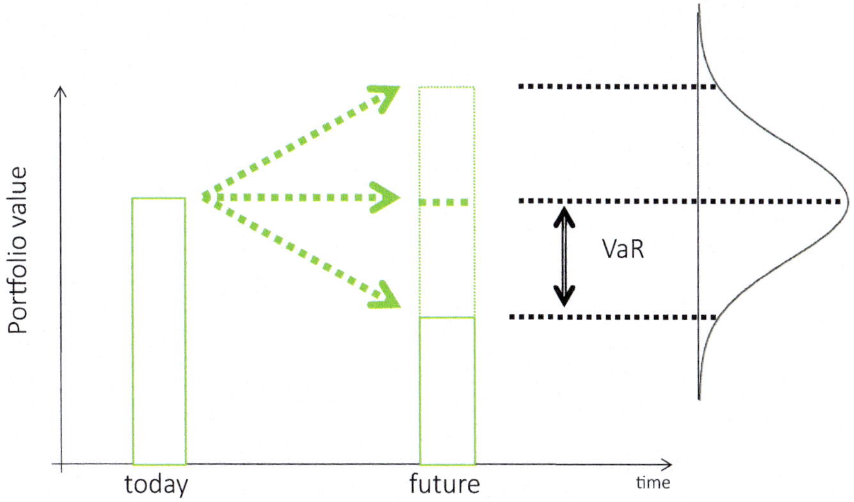

Fig. 4.2 Determination of Value-at-Risk

position (a portfolio) that will not be exceeded within a certain period of time with a certain probability. For example, if the one-day 95% VaR is equal to €100,000, the loss after one trading day will not exceed €100,000 with a probability of 95%.

Figure 4.2 shows schematically how the VaR is calculated. Assume the current portfolio value can develop in three ways in the future: It can rise, stay the same or fall. It is most likely to remain the same (or at not change a lot), which is shown by the continuous probability distribution on the right-hand side of the figure. The higher the possible loss, the less likely it is. The VaR is the loss that will not be exceeded with a certain probability. This value can be calculated in different ways: historical simulation, Monte Carlo simulation and analytical methods. The historical simulations method is very intuitive. Therefore, we will first show how to calculate the one-day 99% VaR this way.

The basic assumption of historical simulation is that future prices will behave similarly to historical prices, i.e. historical observations are representative for future behaviour. Thus, historical information about price developments serves as input, for example from the last 300 trading days. In a first step, this time series of price changes needs to be analysed and it needs to be checked if what happened these days was normal; i.e. if there were no unusual developments such as structural breaks. This requires expert knowledge about the market and should be taken seriously, because if, for instance, the time series is taken from a time with an unusual positive market development, the VaR will most likely be underestimated. Once this is done, daily percentage price changes can be derived. An example of such price changes is shown in Fig. 4.3. Here, the variation from one day to the other is between 2.5%—the largest increase—and −2%, the largest decrease.

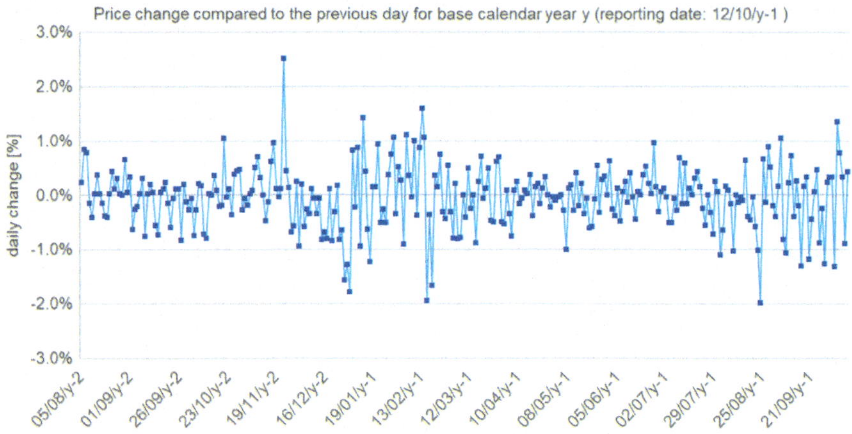

Fig. 4.3 Step 1 to determine the Value-at-Risk using historical simulation

The second step is to apply these historical percentage price changes to the current market price and evaluate the portfolio with the resulting price variations. In other words, the current portfolio value is multiplied with each of these daily price changes and the result is a time series of daily portfolio value changes in euros.

This is shown in Fig. 4.4 for a certain assumed price. The price changes are the ones shown in Fig. 4.3. The figure shows that the positive deviations from one day to the next can amount to around €65,000 and the negative ones to around −€50,000. In risk management, only the worst cases, i.e. the potential losses, are of interest.

In the third step, the price changes are sorted in ascending order as shown in Fig. 4.5. The highest loss of around €50,000 is depicted as the first value on the left, and the highest increase in the portfolio value is on the right. As we are dealing with risk management, our current focus is more on the negative outcomes on the left side.

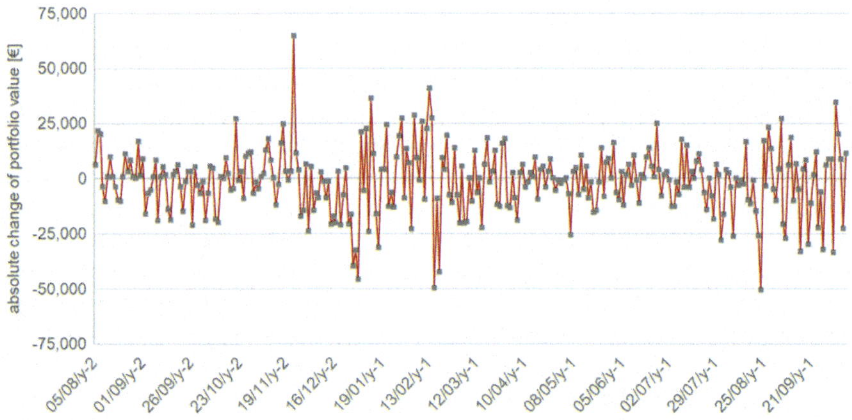

Fig. 4.4 Step 2 to determine the Value-at-Risk using historical simulation

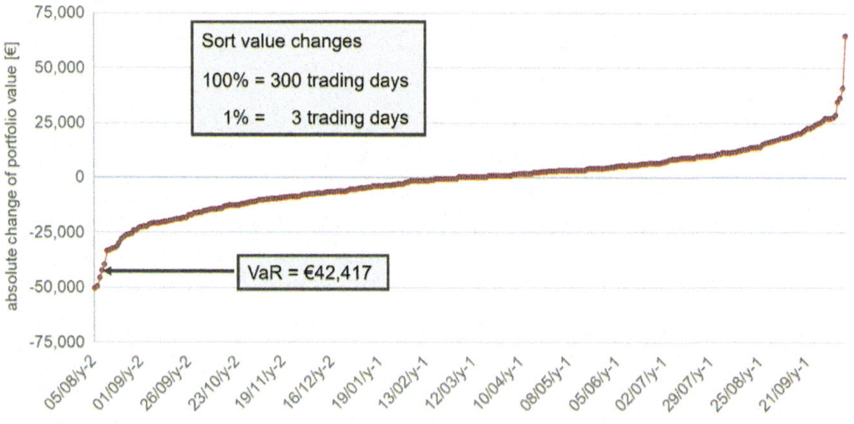

Fig. 4.5 Step 3 to determine the Value-at-Risk using historical simulation

Since the confidence level is 99%, the worst 1% price variations can be discarded. In the case of a period of 300 trading days, these are the three worst values. These are a loss of €50,000 and two other values around €50,000 and €45,000. The fourth worst price variation is then the VaR, which in this example is €42,417. This means that (assuming that the model is correct), the loss from today to tomorrow will not exceed €42,417 with a probability of 99%.

This example shows that the determination of the VaR using historical simulation is rather intuitive. The advantage of this method is that it is relatively accurate (given the assumption that prices in the future will behave similarly to the last 300 days), and it is easier to implement than the Monte Carlo simulation.

The disadvantage of the historical simulation is that an accurate forecast requires a long history. This is a significant data set, particularly because—as mentioned above—only the $x\%$ worst outcomes are of interest for the calculation of the VaR. Furthermore, the implicit assumption that prices in the future will behave like prices in the past tends to become less realistic the further back the data points are. Also, historical simulation is not well suited for abnormal market conditions. As mentioned previously, the calculation of the VaR often assumes normal market conditions by definition.

An alternative to calculate the VaR is via Monte Carlo simulation. Monte Carlo simulation is a method from stochastics where random samples of a distribution are repeatedly drawn. Therefore, to perform a Monte Carlo simulation, first an assumption about the distribution of the portfolio value is needed. This is the distribution we want to draw samples from. A good start here is the Normal Distribution:

$$r \sim \mathcal{N}(\mu, \sigma). \tag{4.2}$$

In this case, the assumption is that the daily returns r of the portfolio are normally distributed with parameters μ and σ, with μ being the mean and σ being the standard

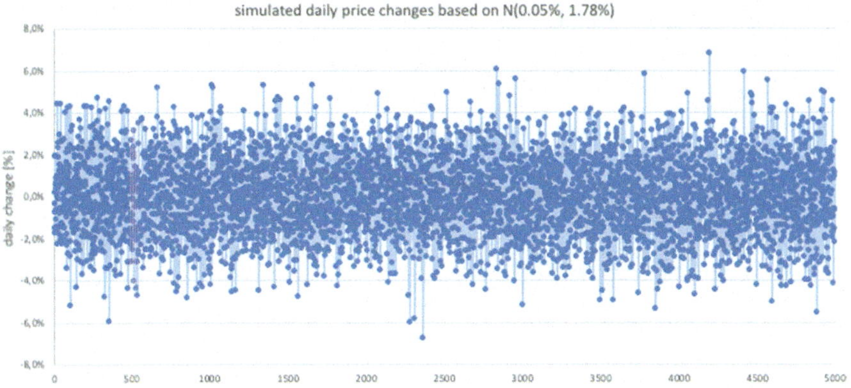

Fig. 4.6 Step 1 to determine the Value-at-Risk using Monte Carlo simulation

deviation. We can, for instance, assume that the average daily portfolio return is 0.05% with a standard deviation of 1.78%. These values can—like in the above example with historical simulation—be estimated from historical daily returns, but they can also include expert knowledge. For instance, if the market was very calm in the past but is expected to become more volatile, the σ parameter may be increased. Also, the assumption on future daily returns of the portfolio may be set below historic observations to be more conservative. This means that the value of μ would need to be decreased.

Once mean and standard deviation are set, in the next step many samples are drawn from this distribution. The more samples are drawn, the more precise the result becomes, but the more computational power is required. This can be done with any spreadsheet software. As a result, you will get many (say, 5000) relative daily changes of the portfolio value. One possible outcome of this simulation is shown in Fig. 4.6.

As with the historical simulation, in a second step these percentage price changes are applied to the current portfolio, which in this example is €109,852,704.[7] The result is shown in Fig. 4.7. This is the distribution of the portfolio value change in euros based on 5000 draws. You can see that the changes range from over −€7 million to over +€7 million.

Again, in risk management one is only interested in the worst outcomes, i.e. in this case in the worst 1% (given a confidence level of 99%). Thus, the portfolio value changes must—like with historical simulation—be sorted in ascending order and the 1% most unfavourable changes need to be discarded. In the case of 5000 simulations these are the 50 worst outcomes. The next value is the Value-at-Risk. This is shown in Fig. 4.8.

The result is a Value-at-Risk of €4.4 million. This means that—given the assumption about the normally distributed price changes is correct—the daily loss of the portfolio will not be higher than that value with a probability of 99%.

[7] The figure is so crooked because this is an actual example from practice.

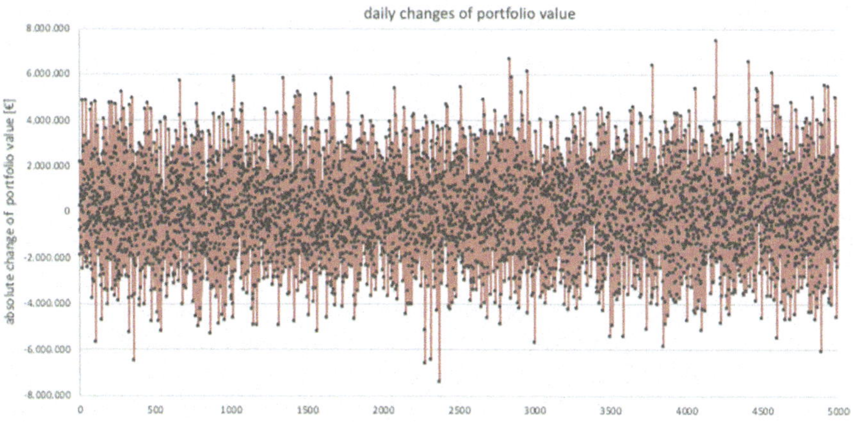

Fig. 4.7 Step 2 to determine the Value-at-Risk using Monte Carlo simulation

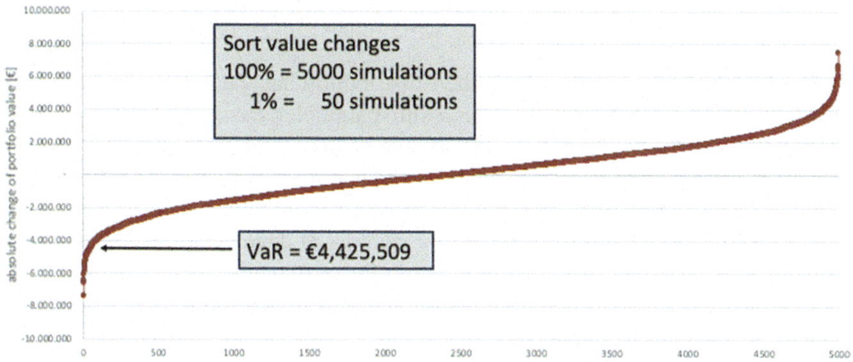

Fig. 4.8 Step 3 to determine the Value-at-Risk using Monte Carlo simulation

The process for calculating the Value-at-Risk using Monte Carlo simulation is similar to historical simulation. However, there is one meaningful difference: in this example, we have assumed that the value changes of the portfolio are normally distributed, with a certain mean and standard deviation, whereas in the historical simulation no explicit parametric assumption was made regarding the distribution of the value changes. The advantage of Monte Carlo simulation is that in Monte Carlo simulation both the parameters (mean and standard deviation) and the distribution itself can be changed.

The change of parameters is useful because the information about the parameters that were estimated from historical values may not match the expectations about the future. As explained above, it can be assumed that the market will become more volatile and thus the standard deviation in the model can be increased. This would ceteris paribus result in a higher Value-at-Risk.

The adjustment of the distribution assumption about the portfolio profit can also be based on expected conditions in the future. In the above example we assumed that the returns are normally distributed. On the one hand, this is a common assumption in finance, mainly because assuming normality of the returns highly simplifies statistical inference.[8] On the other hand, this is problematic since daily returns are typically not normally distributed in reality. It is well known that they can be heavy-tailed, skewed or possessing other kinds of asymmetries. Moreover, they exhibit other properties like volatility clusters or even long memory. These properties are referred to as the stylised facts of empirical finance.[9]

If the risk manager wants to do justice to these properties, a deeper analysis of the price time series is required. For our textbook, we want to show that if required by the risk manager it is possible to choose a distribution assumption other than the normal distribution for the Monte Carlo simulation and thus explicitly model heavy-tailed price returns. We choose Student's *t*-distribution as an example for this in the following. The *t*-distribution can be used to deal with the problem of heavy (or "fat") tails. This means that extreme events (i.e. particularly large or small daily profits or losses) occur much more frequently in practice than is modelled by the normal distribution.

Just like the normal distribution, the *t*-distribution has the two parameters mean and standard deviation. However, it has a third parameter, the *degrees of freedom*. Put simply, the degrees of freedom are a measure of how "thick", "heavy" or "fat" the tails of the distribution are. If the degrees of freedom are very high, the tails are just as flat as in the normal distribution. This means that extreme events are just as rare as in the normal distribution. The lower the number of degrees of freedom, the more probable they become. This is exemplified in Fig. 4.9, which shows the probability density functions of a normal distribution and a *t*-distribution with 5, 10 and 100 degrees of freedom. In all cases the mean μ is set to 0 and the standard deviation σ is set to 1.

Figure 4.9 clearly shows the effect of the degrees of freedom of the *t*-distribution. With 100 degrees of freedom, the shape of the curve (green) is almost identical to the normal distribution (blue). In our example, this means that high daily losses (and profits) are equally unlikely. However, if we change the parameter for the degrees of freedom to 10 (yellow), we see that the tails of the distribution are clearly "fatter"; i.e. high daily losses (and profits) are more likely. This effect is amplified if the parameter for the degrees of freedom is set to 5 (red).

To show the effect of allowing for fat tails in practice, we repeated the previous Value-at-Risk calculation. This time we used a t-distribution with 5 degrees of freedom instead of a normal distribution. The steps remain the same:

---

[8] Actually, in the case of the normal distribution, it would not have been absolutely necessary to calculate the 1% percentile with the Monte Carlo simulation. This is because there are tables with these percentile values for the standard normal distribution, i.e. the normal distribution with mean 0 and standard deviation 1. These values can easily be transformed into any normal distribution with mean μ and standard deviation σ. However, for other distributions such as the t-distribution or other more complex distributions, such tables are not readily available.

[9] Bade et al. (2009).

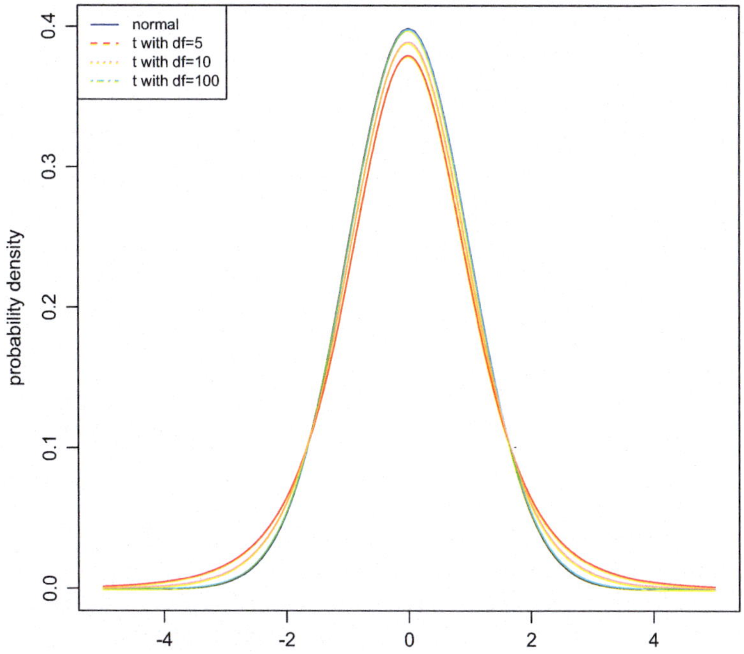

Fig. 4.9 Comparison of a normal distribution with a *t*-distribution with 5, 10 and 100 degrees of freedom

- Step 1: Draw many (in this case 5000) samples from the daily price change distribution.
- Step 2: Apply relative price changes to the portfolio value.
- Step 3: Sort the portfolio price changes in ascending order and discard the 1% most unfavourable value changes (confidence level = 99%). The next most unfavourable value is the Value-at-Risk.

Figure 4.10 shows the outcome of step 3, i.e. the sorted absolute price changes. The same mean value (0.05%) and standard deviation (1.78%) were chosen for this as in the example with the normal distribution. Also, the initial portfolio value is the same (€109,852,704). The only difference is that this time the *t*-distribution with 5 degrees of freedom was chosen to allow for fat tails. The effect is obvious: The Value-at-Risk is significantly higher compared to the above example with the normal distribution, namely almost €6.4 million compared to €4.4 million. This is because extreme events occur much more frequently with the t-distribution with few degrees of freedom.

This example illustrates the impact of the choice of the distribution assumption on the Value-at-Risk using Monte Carlo simulation. Modelling the performance of a portfolio correctly is an important task in the practice of quantitative risk management. Typically, a portfolio contains more than one asset, so more than one price path must be modelled. For example, the portfolio profit may depend on year-ahead

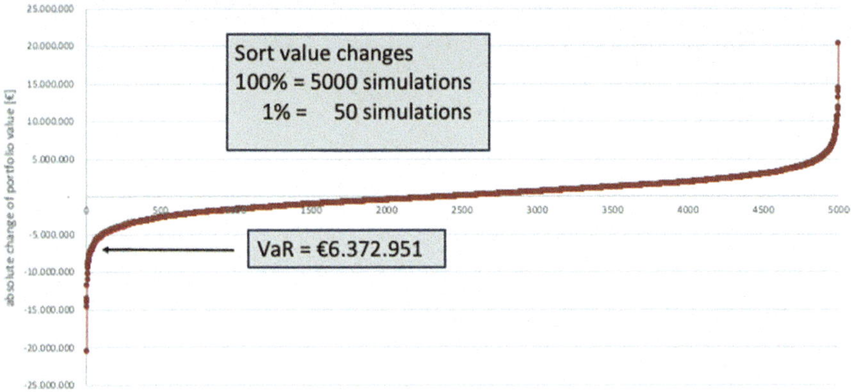

Fig. 4.10 Step 3 to determine the Value-at-Risk using Monte Carlo simulation with *t*-distributed daily profits

power forward and year +2 power forward at the same time. These are highly correlated and therefore need to be modelled via a multivariate distribution, where not only the individual price developments but also their correlation needs to be estimated. Another example can be that a portfolio contains both power and gas futures. These prices are also highly correlated, but probably not as strongly as those of the year-ahead and year +2 electricity futures.

The Monte Carlo approach is flexible enough to deal with multivariate distributions. We have not covered this topic in detail here as it is beyond the scope of this book. If you are interested in this topic in more depth, we refer you to the relevant literature in the field of quantitative finance.[10]

In addition to the historical and Monte Carlo simulation, an analytical determination of the Value-at-Risk is also possible. We have already hinted at this methodology in the Monte Carlo simulation above. If the daily price changes are assumed to be normally distributed, both parameters (mean and standard deviation) and therewith the whole distribution can be approximated from historical data. This is shown in Fig. 4.11, where the purple bars represent a histogram of historical price changes, and the red line is the corresponding normal distribution.

The percentiles of the standard normal distribution (with mean 0 and standard deviation 1) are known, can be found in standard textbooks and can be transformed to any other normal distribution with the so-called z-score. For instance, the z-score of the 99% percentile of the standard normal distribution is 2.326. This means, that if you want to find the 99% percentile of any other normally distributed variable, you have to multiply its standard deviation by 2.326. This makes it pretty straightforward to calculate the 99% VaR of a portfolio with a standard deviation of σ. It is simply the current portfolio value (i.e. price times quantity) multiplied by σ and 2.326, as the following equation shows:

[10] For instance, we suggest Barbu and Zhu (2020).

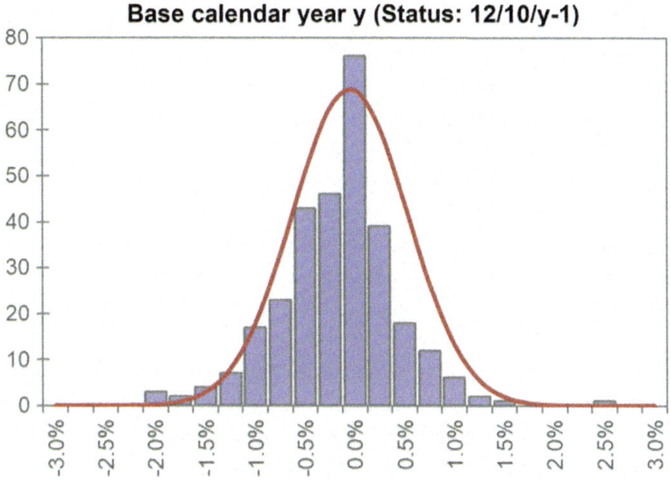

Fig. 4.11 Analytical determination of Value-at-Risk

$$\mathbf{Var\,(99\,\%) = 2.326 \times \sigma \times price \times quantity} \tag{4.3}$$

It must be mentioned that analytical determination of the Value-at-Risk only works under strict assumptions. Typically, analytical methods are only available if daily price changes are assumed to be normally distributed, which is rarely the case in practice.[11] Therefore, applying this methodology should only be done if the portfolio composition is very simple and if it is followed by more sophisticated approaches such as Monte Carlo simulation. However, this method can be used to gain an initial impression of how risky a portfolio is and what the VaR is likely to be.

4.3 Credit Risk Management

Credit risk or counterparty default risk arises as soon as contracts for energy supply or consumption are concluded. There is always a counterparty, and this counterparty can always default. Even with exchanges as counterparty, the probability of default is extremely low due to regulation—but it is not zero.

Credit risk only exists for the creditor. If a debtor owes something to a creditor, the debtor's credit risk is zero. Furthermore, we assume credit risk is always positive, i.e. the debtor does not benefit when the creditor defaults because the creditor's bankruptcy trustee will insist on all payments owed to the bankrupt counterparty.

Credit risk consists of two components:

---

[11] See the above comment on the stylised facts of empirical finance.

- **Resale or Rebuy Risk**: This is the difference between the current market price and the contracted price of undelivered energy. For instance, assume a company sold energy to a counterparty for 55 €/MWh, and after that the price drops to 50 €/MWh. If the counterparty goes bankrupt and cannot take (and pay) the energy, the seller will have to resell it. However, its market value is now 50 €/MWh, and nobody will pay the originally contracted price of 55 €/MWh. The difference is the seller's loss due to resale risk.

  Rebuy risk is similar. Assume that a company bought energy at a low price. After that, prices rise, and its counterparty (the energy deliverer) goes bankrupt. Now that company needs to procure at the current (i.e. higher) market price. The difference is the rebuy risk.[12]
- **Free Delivery Exposure**: This is the value of the energy quantities already delivered but not yet paid for. The supply of energy begins at the start of the delivery period, but payments are generally not made until later. As soon as delivery starts, the free delivery exposure arises and increases until the first payment is made with the delivered energy.

The total credit risk or credit exposure is the sum of both components:

$$\textbf{credit exposure} = \textbf{resale/rebuy risk} + \textbf{free delivery exposure}. \tag{4.4}$$

Resale/rebuy risk always arises as soon as forward prices change. As we showed in Sect. 2.1.1, this can often be the case on energy markets. Regarding free delivery exposure, payments for the delivered energy can occur surprisingly late. As an example, we can see in Fig. 4.12 a screenshot from an EFET contract.[13]

The EFET draft contract specifies in section 12 "invoicing and payment" that the invoice is to be sent "in the course of the calendar month following a delivery". Additionally, it specifies the payment on or before the 20th calendar day of the calendar month following delivery (or the next business day), if the bill is sent timely. This means that for the energy delivered during any given month the payment is due on the 20th of the month after delivery. This is the origin of the free delivery exposure.

If two companies sign an energy delivery contract with these payment conditions, and energy is delivered during January and continues to be delivered during February, then on the 20th of February a first payment is received. However, this payment covers the energy delivered in January only, despite at that point in time energy has already been delivered for 20 days in February as well. Even if the buyer fulfils all his payment obligations, the free delivery exposure is never zero until the final payment.

---

[12] Of course, market prices might also have fallen before the bankruptcy. In this case, the buyer would be interested in not fulfilling the contract and buying elsewhere at a more favourable price. But even in the event of bankruptcy, you are obliged to fulfil the contract, and in this case the seller's bankruptcy trustee will insist that the buyer does so. Therefore, this risk cannot be negative.

[13] The European Federation of Energy Traders (EFET) started when Europe liberalised its energy markets. Their main task was to set up market guidelines and rules, and they offer some standard contracts. If two parties agree to an energy transaction, they can use the EFET standard contract and then adapt it to their needs or simply take it as it is because it is not biased and it saves work for both of the parties.

§ 12
Invoicing and Payment

1 **Invoice:** Each Party who is a Seller of Certificates in an Individual Contract shall transmit to the other Party in the course of the calendar month following a Delivery of Certificates for the previous month an invoice setting forth the total quantities of Certificates that were sold by it under Individual Contracts in the previous calendar month. In connection with such invoice the Party may state all amounts then owed between the Parties pursuant to the Individual Contracts including, without limitation, all amounts owed for the purchase and sale of Certificates, fees, charges, reimbursements, damages, interest, and other payments or credits owed between the Parties and, if applicable, any net amount due for payment pursuant to § 12.3 (*Payment Netting*).

2 **Payment:** On or before the later to occur of (a) the twentieth (20th) calendar day of the calendar month or if not a Business Day the immediately following Business Day or (b) the fifth (5th) Business Day following receipt of an invoice (**"Due Date"**), a Party owing an invoiced amount shall pay, by wire transfer in freely available funds, the amount set forth on such invoice to the payment address or bank account provided by the other Party as specified in the Election Sheet. Unless otherwise specified in the Election Sheet, such payment shall be made in EURO, and subject to § 13 (*VAT and Taxes*), and the remitter shall pay its own bank charges.

3 **Payment Netting:** If this §12.3 is specified as applying in the Election Sheet, if on any day the Parties are each required to pay one or more amounts in the same currency (for which purpose all EURO currencies shall be considered a single currency) under one or more Individual Contracts, then such amounts with respect to each Party shall be aggregated and the Parties shall discharge their respective payment obligations through netting, in which case the Party (if any) owing the greater aggregate amount shall pay the other Party the difference between the amounts owed.

Fig. 4.12 EFET payment conditions[14]

4.3.1 Credit Risk Example

In the following example, the effects of rebuy/resale risk and free delivery exposure will be demonstrated using typical market data of an energy supply contract with delivery between 01/01/y and 30/04/y. The payment is done according to the EFET contract shown in Fig. 4.12, i.e. on the 20th calendar day of the calendar month following delivery. The parameters of the energy supply contract are as follows:

- Contract signature on 10/10/y-1.
- Delivery period: 01/01/y–30/04/y.
- Base contract of 50 MW sold for 105 €/MWh.
- Current market price: 100 €/MWh.

The timeline of Fig. 4.13 shows the delivery and payment dates of the contract. Contract signature took place on the 10th of October in the year before delivery for a price of 105 €/MWh for 50 MW baseload in the delivery period of January to April. On the 1st of January, the delivery starts, but again, there is no payment in January. The first payment is due on the 20th of February, and delivery continues until the 30th of April. The last payment is due on the 20th of May and is the payment for the energy delivered in April.

In this example, both risk components will be calculated directly before the first payment on 20/02/y (valuation date) from the perspective of the company selling the energy. The resale risk on the 20th of February is the financial risk if the counterparty goes bankrupt and cannot take the energy. In that case, it needs to be resold somewhere

[14] EFET (2020).

Fig. 4.13 Timeline of the credit risk example

else. This needs to be done at the current market price of 100 €/MWh and not 105 €/MWh (the contract price). The amount of undelivered energy corresponds to delivery from the 20th of February to the 30th of April, i.e. the 8 remaining days of February, 31 days in March and 30 days in April. This is a total of 69 days. The amount is 50 MW for each of the 24 hours of the day,[15] and there is a loss of €5 for each MWh. This leads to the calculation of the resale risk[16]:

Resale risk: $69 \text{ days} \times 50 \text{ MW} \times 24 \text{ h/day} \times 5 \text{ €/MWh} = \text{€}414{,}000.$

The second component is the free delivery exposure. This is the value of what has already been delivered but not yet been paid. The seller delivered for 31 days in January and 20 days in February. This is a total of 51 days. In all these days they delivered 50 MW every hour of the day for a contract price of 105 €/MWh. Therefore, the free delivery exposure is the following:

Free delivery exposure: $51 \text{ days} \times 50 \text{ MW} \times 24 \text{ h/day} \times 105 \text{ €/MWh} = \text{€}6{,}426{,}000.$

The credit risk is the sum of both components:

Credit risk: $\text{€}414{,}000 + \text{€}6{,}426{,}000 = \text{€}6{,}840{,}000.$

This is the financial damage in the event of the bankruptcy of the counterparty on the 20th of February before any payment has been made. Of course, this is the worst-case scenario because this happens just before an expected payment. However, this bankruptcy is detrimental, regardless of when it happens.

It is not uncommon that bankruptcies are made public just when a major payment is due. This is because the people at the company that goes bankrupt may have been trying to save the company and are working very hard to find a solution to their financial struggles, like, e.g. another big deal or a bank loan. Then at one point in

[15] Strictly speaking, depending on the country, March has one hour less due to the time change from summer to winter time. However, for the sake of simplicity, we will not consider this here.

[16] When calculating this in real life, we recommend checking the units. This will help you to know if everything was done correctly or an element of the calculation is missing. Here we can verify that we did everything right because the risk is the potential loss, and this is quantified in euros.

time, there is some bill that needs to be paid, and it is at that point that the company realises that they cannot pay it and declare bankruptcy.[17]

As soon as a payment has not been made, from the seller's point of view the energy supply should be stopped immediately. Some may argue that this is not customary. Indeed, in practice energy delivery may continue. Imagine that in the above example on the 20th of February the customer does not pay. The back office of the supplier realises that immediately and calls them to ask what is going on. The customer may tell them that they are in financial problems but that they are working on it. Since there is an established relationship with that client, the supplier may continue to deliver for a few more days. This may be more customary, but it can also mean that even more money is lost, because if the customer is bankrupt, it is very likely that this energy will not get paid either. In the above example, the free delivery exposure increases by $50\text{ MW} \times 24\text{ h/day} \times 105\text{ €/MWh} = €126{,}000$ on each day on which energy is delivered and not paid for. This needs to be kept in mind.

As said before, this is the worst-case scenario. Often, when a company goes bankrupt, this does not mean that they are not paying their obligations at all. There is typically a lawyer responsible for the insolvent company, who will try to get as much revenue as possible to pay as much debt as possible. So, in the above example, the supplier may still receive a part of the €6,840,000. But still, the order of magnitude and the importance of properly managed credit risk is evident.

Figure 4.14 shows the development of credit risk in the example, split into the two components rebuy/resale risk and free delivery exposure. The red line represents the rebuy/resale risk component, and the blue line is the free delivery exposure. All numbers are shown in monetary units (Euros, vertical axis). The horizontal axis depicts time.

Time starts on the 10th of October with the conclusion of the contract, and rebuy/resell risk may appear immediately, due to the fluctuations in the forward price. Again, neither of the two risk components can become negative (i.e. become a chance), but they may be zero.

In this example, the price declines towards the end of the year before delivery, and the more it falls below 105 €/MWh, the greater the resale risk becomes. The red line is thus going up and reaches its peak in the example before delivery starts. This is typical because a customer default before the 1st of January poses resale risk for the full four months of delivery, i.e. the maximal amount of energy.[18] Once delivery starts, resale/rebuy risk declines because the undelivered volumes decline. However, resale/rebuy risk remains to be influenced by the price development. Hence, price developments after delivery begin still increase or decrease this risk component. Nonetheless, it will decline as time goes by and reach zero at the end of the delivery.

With the start of delivery, the second component of credit risk, the free delivery exposure, starts to grow and reaches its peak just before the first payment. As the

---

[17] It should not be like this in practice. Companies with serious financial problems should be more proactive and may get in legal trouble if they wait too long to declare bankruptcy.

[18] In the previous calculation, instead of multiplying by 69 days, the supplier would have to multiply by 120 days.

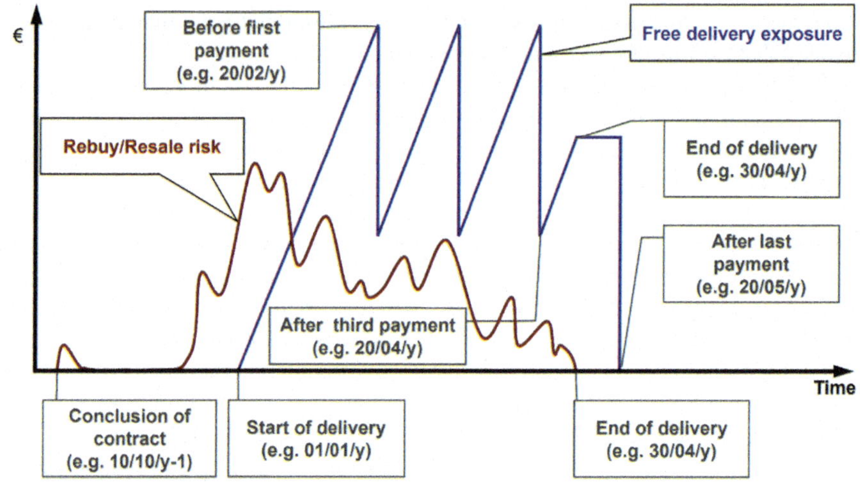

Fig. 4.14 Variation in time of the components of credit risk

figure shows, the first peak is on the 20th of February. After the first payment which is due on that day, it drops considerably but does not go to zero, because the payment received in February covers the energy delivered in January only. Therefore, energy for 20 days of February has already been delivered but will not be paid until the 20th of March.

With additional deliveries, the free delivery exposure starts to grow again until the 20th of March, when all electricity that has been delivered during February is paid. This cycle repeats until April. After the payment on the 20th of April, the free delivery exposure starts to grow again, but it stops growing at the end of delivery on the 30th of April, because there is no more electricity being delivered after that day. However, the last payment is received on the 20th of May. At that date, the contract is financially concluded, and its credit risk is zero.

4.3.2 Factors Influencing Expected Credit Risk

The above example showed how a straightforward delivery contract translates into credit risk. Both resale/rebuy risk and free delivery exposure change after the contract has been concluded as market prices change (resale/rebuy risk) and delivery begins (free delivery exposure). While this is true, the example contained a slight simplification in the calculation of the daily loss associated with a counterparty default.

Credit risk management in reality is even more complex as the creditor may be interested in additional factors. First, a credit risk assessment is needed before a contract is signed, e.g. to quantify the risk capital required for the contract. This can be perceived as an expected credit risk.

The expected credit risk is influenced by several factors:

- **Contract Volume**: The higher the contract volume, the higher the expected credit risk. For instance, a 100 MW contract implies more credit risk than a 1 MW contract. A 1 MW contract for one year implies more credit risk than a 1 MW contract for one month. Note that contracted volumes in energy trading are often significant. For example, a 5 MW front-year contract at 100 €/MWh has a financial volume of €4.38 million and the associated expected credit risk is substantial.
- **Price Volatility**: The more volatile prices are the higher the expected credit risk will be because a significant resale/rebuy risk is more likely to manifest after the contract has been signed. Forward electricity prices have been rather volatile and associated expected credit risk is substantial.
- **Time Between Conclusion of the Contract and Start of the Delivery Period**: Expected credit risk increases with the time between the conclusion of the contract and the start of the delivery period. For instance, if a contract is concluded with delivery 10 years from now, a lot can happen until delivery starts: Prices are more likely to fluctuate strongly, and the position can get very valuable. On the other hand, for a contract with delivery next month, the probability of large price changes is significantly less likely. Credit risk increases with a price increase after a purchase and a price decrease after a sale. In other words, whenever the price moves in a favourable direction, the contract becomes more valuable and that is mirrored one-to-one in the credit risk. Forward contracts in electricity often have a comparably long time between conclusion of the contract and start of delivery. Again, associated expected credit risks are substantial.
- **Number of Opposing Positions Without Netting Agreement**: Expected credit risk with a counterparty increases with the number of opposing positions without netting agreements, as explained above.
- **Creditworthiness of the Counterparty**: Expected credit risk is negatively correlated with the creditworthiness of the counterparty. If a counterparty has low creditworthiness, then it is more likely to go bankrupt and not pay its bills. Doing business with a government-controlled municipal utility with low default probability is not the same as doing business with a start-up energy trading company.

As this list shows, credit risks in energy trading are manifold and substantial. They need to be monitored and managed carefully. While a company cannot influence most of them (e.g. the price volatility of a certain contract or the creditworthiness of an external company), it can manage its engagement with both, i.e. decide whether and to what extend sign a contract with either. Companies thus need to determine how much money they will allow a counterparty to owe them, and they should base this decision on the counterparties' creditworthiness. Even though there is no guarantee that a debtor with a high creditworthiness will be able to pay its bills, it is more likely.

4.3.3 *Credit Rating and Credit Limits*

The assessment of a counterparty's creditworthiness and trustworthiness is an important yardstick to determine how many contracts and how much expected credit risk you are prepared to sign with that counterparty. One of the main problems with default events is that they may—or may not—happen in the future and it is not possible to know in advance how a counterparty's solvency will develop in the future. However, there are ways to estimate a company's creditworthiness. These can be divided into three options:

- **Quantitative Analysis**: This means that the capital structure of a counterparty is evaluated. This includes the development of its business in general, as well as the company's liquidity and profitability.[19] A starting point for quantitative analysis is the counterparties' balance sheets, but additional information (both free of charge and commercially available databases) should be used. Typically, energy trading companies have one or more in-house experts analysing all these factors and estimating the preliminary credit ratings of their counterparties.
- **Use the Results of an External Rating Agency**: There are companies specialised in providing credit ratings and the default risk of counterparties. Examples for rating agencies are Standard & Poor's, Moody's, or Fitch Ratings. On the one hand, these companies are specialised on company ratings. They have dozens of experts and high knowledge in their area of specialisation. Furthermore, it is economically efficient if, e.g. a large trading company is rated only once by a specialised rating agency instead of hundreds of times by individual trading partners. On the other hand, companies using external ratings rely on the external experts and have to trust their assessment.
- **Qualitative Analysis**: This can be used if neither quantitative information nor results from external rating agencies are available (or if they are too expensive). A qualitative analysis is based on the type of company, its legal form, its stock market listing and the development of the sector.[20] The outcome is a preliminary credit rating for a contractual partner. While valuable, it is usually inferior to quantitative analyses.

Sometimes the credit rating of a parent company must be assessed. For instance, a top-tier global oil company may set up a small trading company as a separate legal entity. As a consequence, it is hard to gain information on the daughter company (small, potentially new, not stock-listed, brand is established by parent company, etc.). Essentially, potential business partners have to consider the likelihood that the oil company will bail out the small trading company if something goes wrong. From

[19] Profitability is an intuitive sign of solvency. If a company has been very profitable over the last ten years, they are likely to be able to pay their bills this year.

[20] If no information about the company is available, the development of the sector is sometimes used as a proxy: If for example everyone in the sector earned high revenues last year, it may be reasonable to assume that the company in question is not the single exception.

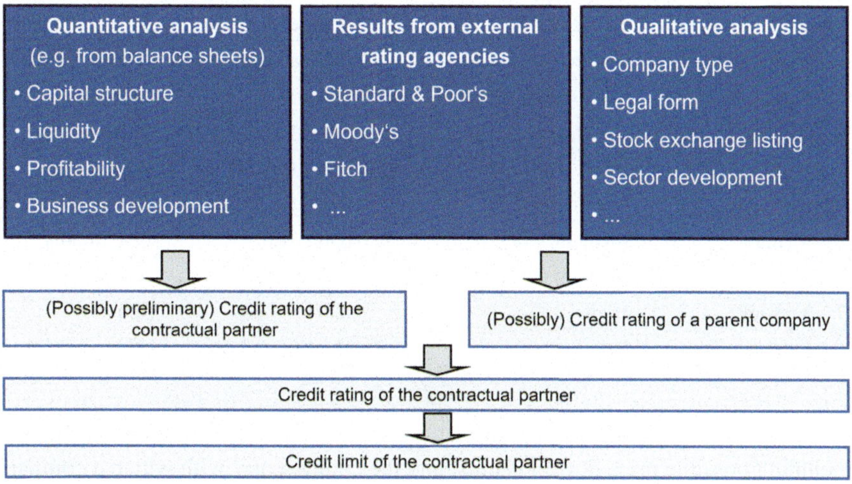

Fig. 4.15 Overview of the credit rating of contractual partners

a legal point of view, they may not have to. However, it is not unlikely that they would do so anyway to avoid a damage to their reputation.

Once the default likelihood of a contractual partner is determined, a credit limit for that company needs to be specified: there needs to be an assessment on how much expected credit risk can be signed with a counterparty based on its credit rating. If the credit rating is low, trading with that company can be prohibited. This should not happen too often, otherwise there may be no one left to trade with. If a counterparty has a high credit rating, it is very unlikely to default. As a result, a company may decide to sign contracts up to €10 million of outstanding expected credit risk with that counterparty. Once that limit is reached, they need to take preventive measures.

The process to determine the credit limit is shown in Fig. 4.15.

The result of the credit rating process is a credit limit for each counterparty. Table 4.1 shows an example of this for a specific company. There are eight different credit ratings in the leftmost column, the associated counterparty default risk, the credit rating number (which in this example goes from + 6, which is the best, to − 6, the worst), and on the rightmost column, there is the respective credit limit.

The first line after the header is credit rating 1 "extremely high", and the counterparty default risk is almost zero. The credit rating number associated with this in the quantitative system is between 4.5 and 6, and this results in a credit limit of €50 million. In other words, if a counterparty is categorised in this class, the company allows €50 million of expected credit risk with this counterparty. If the credit limit is "very high" there is still a credit limit of €30 million. Further down in the table, we see that with a "low" credit rating, traders of the company are only allowed to have an outstanding credit of €250,000 with a respective counterparty. With credit ratings 7 and 8, "very low" and "extremely low", the counterparty default risk is high and very high, and the company's traders are not allowed to trade with them at all.

Table 4.1 Allocation of the credit risk limits per trading partner depending on the credit rating

Credit rating	Counterparty default risk	Credit rating number	Credit limit
1—Extremely high	Almost zero	6.0 to 4.5	€50 million
2—Very high	Very low	4.5 to 3.5	€30 million
3—High	Low	3.5 to 1.0	€15 million
4—High to medium	Low to medium	1.0 to 0.0	€7 million
5—Medium	Medium	0.0 to − 1.0	€3 million
6—Low	Medium to high	− 1.0 to − 2.0	€0.25 million
7—Very low	High	− 2.0 to − 4.5	€0
8—Extremely low	Very high	− 4.5 to − 6.0	€0

To summarise, managing credit risk requires knowledge of both the market, including possible price developments, and the counterparty with which a company is trading. The above analysis, including a hands-on example, has shown that credit risk can be significant even for a simple contract with a term of only a few months. The credit risk can be even higher, especially in volatile markets and for energy supply contracts with terms extending far into the future. Credit risks arising from the bankruptcy of a counterparty can lead to serious financial difficulties. Their management is therefore crucial for any successful trading business.

4.4 Product Liquidity Risk Management

We have already introduced the concept of liquidity (see in particular Sect. 2.6 and the brief description of product liquidity risk at the beginning of this chapter). Remember that standardised trading products (i.e. futures and forwards on electricity and natural gas, but also shares in general) are traded on continuous trading markets. As a rule, both buyers and sellers are free to participate in trading. Accordingly, a buyer can only buy if a seller wants to sell the corresponding quantities at the same time—and vice versa. In addition, both must agree on a price. In that sense, liquidity risk manifests whenever the market "dries up" because buyers and sellers refrain from engaging in the market. Hence, we will describe this process in more detail in the following.

Assume a company is long but needs to reduce its position, e.g. for reasons of price risk management. If bids are available on the market, this potential seller has two options: Its trader can either post an offer himself (and thus signal his interest in selling to other market participants) or accept an existing bid. The first option is usually chosen when the seller wants to sell at a price above the highest currently available bid. The second option is attractive if the seller needs to act quickly, as the deal can be finalised immediately.

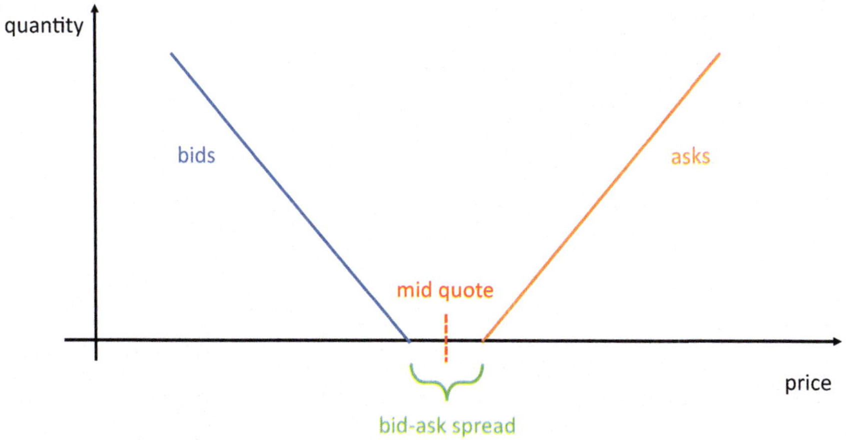

Fig. 4.16 Typical shape of an order book

This general setting also clarifies that the cheapest offer is always higher than the highest bid, because otherwise a potential buyer (the actor behind the bid) would be willing to pay more than a potential seller (the actor behind the offer) demands. These two parties could and should engage immediately in a trade beneficial for both parties. The difference between the lowest offer and the highest bid at any point in time is also called the bid-ask spread or bid-offer spread at that point in time. A market can have different numbers of bids at different prices on both the bid and the offer side. The totality of all bids and offers on the market at a point in time form the so-called order book at that moment. We explained order books and the bid-ask spread in detail in Sect. 2.4.2.

Figure 4.16 shows schematically what an order book looks like. On the left are the cumulative bids, on the right the cumulative asks. The difference between the highest bid and the lowest ask is the "bid-ask spread", and the midpoint in between is the "mid-quote" or "mid-price" sometimes published by exchanges.

Based on the previously introduced definition of liquidity in Sect. 2.6, we can differentiate five characteristics of a liquid market for a product: (i) tightness; (ii) immediacy; (iii) depth; (iv) breadth; and (v) resiliency.[21] Product liquidity risk thus subsumes all risks caused by changes in these five characteristics. The risks from the first are clear at this point in our book: a widening bid-ask-spread (tightness) can cause costs when a position is closed. This is a typical product liquidity risk. Additionally, the order book (and associated characteristics depth and breadth) also contains highly relevant risks, especially when a company has large open positions which need to be closed fast.

Risks in the order book can manifest in various ways. When closing the position, the company would first trade the best available bid or offer. However, the quantity associated with that position is constrained, potentially leading to a situation where

[21] See Sect. 2.6 for a more detailed definition of these characteristics.

Bid/ask overview

Bid Vol	Bid	Ask	Ask Vol
626	4.45	4.455	1,102
103	4.44	4.46	854
1,411	4.435	4.465	1,908
3,585	4.43	4.47	5,439
4,514	4.425	4.475	3,531
2,999	4.42	4.48	4,645
4,544	4.415	4.485	3,667
10,228	4.41	4.49	2,538
3,056	4.405	4.495	2,698
3,484	4.40	4.50	1,785

Prices and limits are from 28/08/23 15:35:55

Fig. 4.17 Xetra order book Borussia Dortmund, 28.08.2023 15:35:55[22]

the desired trading volumes surpass the available quantities at the best bid or offer. Most often, the order book comprises additional quantities, albeit at less favourable conditions. The risk is thus that the company trades at rapidly worsening prices—even if volumes in the order book are sufficiently high. Furthermore, there may not be enough volume in the order book to trade desired volumes quickly. In such instances, the company is compelled to wait, exposing itself to other risks, such as price risk.

In the following, we discuss an example of product liquidity risk using the share price of a well-known German soccer club: Borussia Dortmund. Figure 4.17 uses publicly available share information to show how an order book can look like. It is based on original data from the Frankfurt Stock Exchange. The example shows the ten best bid and ask limits with volume for the Borussia Dortmund share listed in the S-Dax on 28.08.2023 at 15:35:55. The highest bid, i.e. the highest price at which a buyer was willing to buy shares of Borussia Dortmund at that time, was €4.450. The lowest offer (referred to here synonymously as "ask") was €4.455. This is the lowest price at which a seller would sell a share in Borussia Dortmund at that time. The bid-offer spread was therefore €0.005.

[22] Börse Frankfurt (2023).

However, Fig. 4.17 contains further information. In particular, it shows what quantities could have been bought or sold at the respective prices. For example, at the price of €4.450, a sale of a maximum of 626 shares is possible (recognisable in the "Bid Vol" column).

Let us now assume that a seller wants to quickly sell a larger quantity of shares, for example 20,000 shares. If the seller could sell all of these at a price of €4.450 (the current highest bid), this would result in sales revenues of €89,000.00 (20,000 × €4.450). However, this is not possible with the given order book, as there is only demand for 626 shares at this price. After that, in the case of an unconditional sell order, the other bids on the left side of the order book in the figure would be filled. Specifically, the seller would sell 103 shares for €4.44, the next 1411 shares at €4.435, then 3585 shares at €4.43, and so on until the last 2218 shares would be sold at a price of €4.41. The total realised revenues for 20,000 shares in this example would be €88,455.525. This amount is approx. 0.61% lower than assuming a sale at the bid price for €89,000.

Hence, the finite depth and breadth in the order book add cost whenever large positions are traded. In terms of liquidity risk, these uncertain variables could develop negatively and increase this cost. Furthermore, the whole market for the product may "dry up", effectively preventing closing the position at all.

Transferring this example to energy trading, we can summarise that there are considerable risks associated with closing (or opening) a position. The size of the effect depends on the quantity of the contracts to be traded (the smaller the quantity, the smaller the discount) and on the shape of the order book at the time of the transaction (the more liquid a contract is traded, the smaller the discount). The latter is the reason why energy trading companies often limit trading of futures contracts to maturity, e.g. up to three years ahead–but not further into the future as liquidity will be lower for such contracts.

Table 4.2 shows an example of the traded volumes in the baseload power contract in the German market area for the next six years. The traded volumes for front year ($Y + 1$) base contract on that day was 1,264,896 MWh, which corresponds to 144 MW baseload. Traded volumes decrease the further the delivery year is in the future. For $Y + 2$ it was 341,640 MWh (i.e. 39 MW), for $Y + 3$ it was only 96,360 MWh (11 MW), and for $Y + 4$ and all consecutive years, the trading volume was zero, i.e. nothing was traded. This means that if a company wanted to close an open position with baseload power with delivery more than three years from now, it would have been challenging and probably not possible at the published settlement price (which is an estimation published by the exchange when no volumes are traded)

If a sale (or purchase) of large quantities would lead to expected price discounts (or price premiums), market players often pre-empt this in the agreed terms of the transaction. Frequently, such transactions are then not executed through the exchange (or the order books of a trading venue described above), but bilaterally, and thus without directly influencing the market price. Such a sale of a larger quantity of futures contracts is also referred to as a "package sale" or "block trade". In this process, a price premium or discount is often agreed: "The package discount is the discount that sellers of a package of shares grant to the buyer. A package is said to

Table 4.2 Trading volume base load German power futures[23]

Delivery year	Volume
Y +1	1,264,896
Y +2	341,640
Y +3	96,360
Y +4	0
Y +5	0
Y +6	0

exist when large quantities of shares are owned by an institution. Trading in such parcels is referred to as parcel trading. [. . .] A package discount is mostly granted if the seller wishes to sell the shares for a variety of reasons. The package discount is understood as a price discount, at the current market price of the stock."[24] While these definitions relate to trading in stocks, they also describe the situation in energy markets.

4.5 Risk Management Processes

Previous sections have explained the importance of risk management in energy trading, what the relevant risks are, how they can be managed, and which parameters are most relevant. However, the operational process of risk management needs to be embedded in procedures and frameworks within a firm. For instance, we mentioned above that once the portfolio loss gets close to the risk capital, measures such as closing certain positions need to be taken. Which measures—and who is responsible to act and oversee them—are usually implemented in forms of risk management processes which are documented in a risk manual. In this section we will introduce such a process by explaining how to implement risk management in an energy trading company from scratch.

Implementing risk management within a trading company starts with an initialisation period. At this point, we assume that nothing has been done regarding risk management within this company yet. This could, for example, be because the company operates in a recently liberalised market and has only just started trading. This assumption also has didactical advantages as it allows to introduce the process step by step.

The first step for implementing a risk management procedure in a company is to analyse the company's organisational chart, which shows the different functions within the organisation. With this, you can get a first impression of who is responsible for what. Employees in areas such as procurement, sales or controlling are of particular importance in our context, as their activities are all related to the open position.

[23] Source: Own illustration based on data published by EEX.

[24] Own translation, based on Börsennews (2023).

Hence, it is necessary to analyse the rights and obligations of those involved in the process of procurement and sales. It is also necessary to examine how risk capital is determined and allocated.

Next, all processes that can affect the open position need to be reviewed and it must be made sure that organisational guidelines exist. This covers the employees as well as the systems that are used for procurement, sales and trading.

In the procurement area, this review looks at the framework of contracts, i.e. whether the company uses the EFET format and their guidelines.[25] It is also important to know the current contractual situation of the company, i.e. the portfolio of signed contracts. This information also contains how much has been bought and sold. The company's current open position can thus be calculated. At this point in our book, the necessity to determine the open position may sound obvious, but in practice many companies are not even aware that they are exposed to price risk from an open position, not to mention how high it is. Turning to the company's counterparties, the starting point is to determine whether rules for the selection of counterparties are in place and whether all current counterparties comply with these rules. In the details, there are other aspects such as whether counterparties' contracts are flexible or fixed volume contracts, how much flexibility they have, etc.

In the sales area, the analysis typically starts with an examination of the client structure. This involves the number of households and larger, real-time metered customers and whether the company has interruptible loads or counterparties with a low creditworthiness. It is also important to determine whether it sells energy only in one specific region, i.e. has a traditional sales territory (typical for municipalities), or whether it competes with other companies in areas where it has not historically been present. This for example is relevant to assess the probability of customers leaving the company and thus posing a volume risk. Other aspects can be whether the company differentiates between several sales portfolios or uses only one aggregated portfolio.

All these steps are a vital part of the initialisation process, because before implementing any processes or measures, the knowledge of the company's current situation is a prerequisite. Figure 4.18 shows an example of an organisational chart of a trading company. Here you can see the trading team on the left, and the sales team on the right, and between them are the central functions that are used by both.

On the trading side, there are typically different departments, like electricity and gas trading. Within electricity trading there is a front office, a portfolio management team (sometimes differentiating between generation and consumption) and an analysis team. Depending on the size of the company, these can be different persons or even different departments. For gas trading, there may be a similar or slightly different structure depending on the focus of the company. Depending on the size of the company, the region and the commodities traded, there may be additional tasks, for example trading desks for emission certificates, weather derivatives, or hydrogen.

Other central functions of portfolio and risk management are bundled in the middle office, which is usually in charge of modelling, valuation, risk analysis and risk controlling/management. Particularly the latter, risk controlling and management,

[25] If the company is not located in Europe, this can be another standardised contract, obviously.

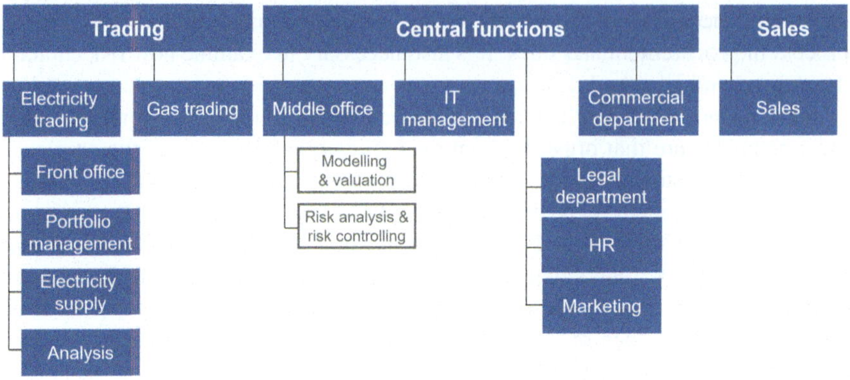

Fig. 4.18 Example of an organisational chart

are (and should) mostly be performed by the middle office, i.e. separated from the front office. This is because if it were in the front office, the actors responsible for trading would control themselves. Risk management should be independent, as traders tend to underestimate their risks and overestimate their opportunities. This is exacerbated by their financial incentives, which often depend on their trading profits in the form of bonuses. Another vital support function is provided by the back office, which is responsible for the smooth functioning of commercial operations, such as record keeping and settlement, trade confirmations, data management, etc. The back office is vital for trading activities because it often controls whether deals were entered correctly into the system by traders and is responsible for cash flow management, including paying the bills. Lastly, IT management, the legal department, HR and marketing are often included in central functions, typically neither part of the middle nor the back office.

Sales activities are often grouped in their own department. Sales is responsible for engaging final customers to buy electricity, natural gas or other products the company is selling. Note however, that different companies have different structures and there is more grey than black and white in this area. For example, sales teams specialised on complex products could be part of the front or the middle office, depending on whether a direct connection to the wholesale market (front office) or to complex option evaluation (middle office) is needed.

Every employee of a company belongs to one (or more than one) of these "boxes". Having such an organisational chart helps to structure employees' responsibilities. For example, someone in gas trading is responsible for trading natural gas—and therefore not for trading electricity or sales activities.

After the organisational structure has been updated and optimised, the next step is to determine the risk capital. Every business activity involves risks, but the risks taken must not jeopardise the continued existence of the company. Risks in energy trading can be substantial, but the risk capital must at least be sufficient to ensure the continued existence of the company even in the worst case.

Since there are different business areas within the company that require risk capital, the total risk capital must be divided between them. This is a strategic decision and depends on strategic importance or expected returns on risk capital, for instance. Risk capital is a necessary prerequisite to enter potentially profitable open positions, but there is only a limited amount of it, and every business unit in the company performing activities that involve risks must have some risk capital for the eventuality of losses. Therefore, anybody in the company who wants to pursue any economic activity has to lobby for risk capital.

Electricity traders, for example, want more risk capital because it allows them higher open positions, potentially higher revenues for the company and bigger bonuses for themselves. The same logic applies to a gas trader, who also needs risk capital to carry out his business. Also, the sales department might want to do business with potential new clients, because this increases their sales volume and their bonuses. However, doing business with new clients involves credit risk and hence requires the assignment of risk capital as well. Consequently, they compete for risk capital with the traders.

The assignment of risk capital can be based on strategic decisions and expected returns. Strategic decisions determine whether the company wants to focus on electricity or gas trading, for instance. A municipality might want to focus on its local core area or expand to other cities or regions. Once this decision is taken, the company can allocate the risk capital according to its strategy.

A focus on expected returns of risk capital maximises revenues regardless of where they come from, i.e. whether they come from trading gas or electricity, within the same region or somewhere else, etc. In such a case it is required to calculate how much risk capital is needed for a deal, and then the risk capital is allocated to where the expected return is maximised.

Figure 4.19 shows an example of a risk capital allocation. Starting from the top, risk capital is first split between credit risk, price risk and other risks. Second, these categories can be divided further. Credit risk capital is allocated to individual counterparties and contracts depending on credit ratings. For price risk, there are loss limits for electricity and gas trading and sales.

Once the risk capital is allocated, the respective limits need to be supervised. This is done monitoring all the different positions in the day-to-day operations.

The framework for risk management, which has been developed so far, is written down in a so-called risk manual. A risk manual defines the different actors' rights and obligations within the company. For example, it specifies in which market a trader is allowed to trade and which products he is allowed to buy and sell. Doing so, it specifies if traders e.g. based in Germany are allowed to buy and sell futures in France or Poland, or if they are allowed to buy and sell hard coal or electricity or hydrogen or natural gas. Typically, within a company there are several different desks, e.g. a coal desk and a CO_2 desk. The people working at a specific desk trade a specific commodity and are not allowed to do otherwise.[26]

[26] These responsibilities and authorisations should also be implemented in the company's systems, i.e. a specific trader should only be able to trade what he is allowed to trade.

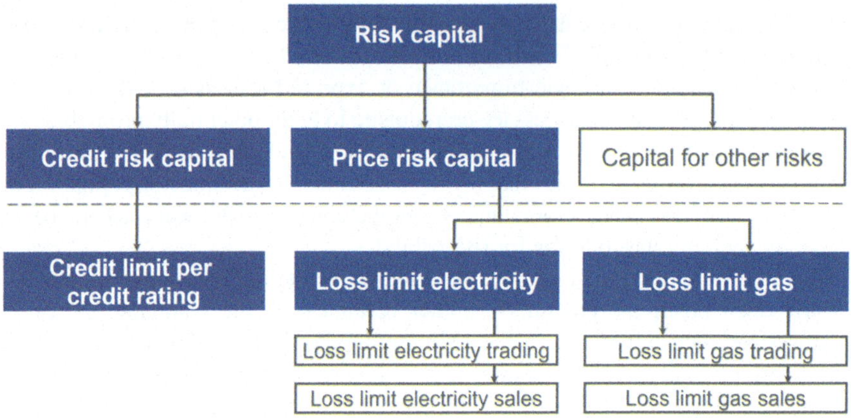

Fig. 4.19 Example of risk capital allocation

Changes in the environment require adjustments in the risk manual, so it needs to be revised periodically. To summarise, the risk manual must include the following elements:

- Allocation of rights and obligations.
- The risks involved and their description.
- Representation of processes and interfaces.
- Description of mandates (limits, methods, products, markets).
- Procedures in exceptional situations (for instance if the managing director is responsible for certain decisions but cannot be reached).
- Description of a portfolio structure.

As explained above, as soon as risk capital is allocated, all positions affecting the risk capital must be monitored in day-to-day operations. This is referred to as risk reporting. Risk reporting needs to be done at least daily and consists of several reports that include all important performance and risk indicators that are needed to determine the company's current performance and risk situation.

A full risk report will usually contain the current portfolio value. Also, it should contain the previous day's trading activities, i.e. what has been bought and sold and how did that affect the open position. Also, the Value-at-Risk and other risk indicators are usually part of most risk management reports. They are usually created overnight and then distributed to responsible employees within the firm in the morning. In addition to daily reports, there are often also weekly, monthly, quarterly or annual risk reports. These summarise the performance and risk development over a longer period of time and serve as a basis for decision-making by the management or a risk committee.

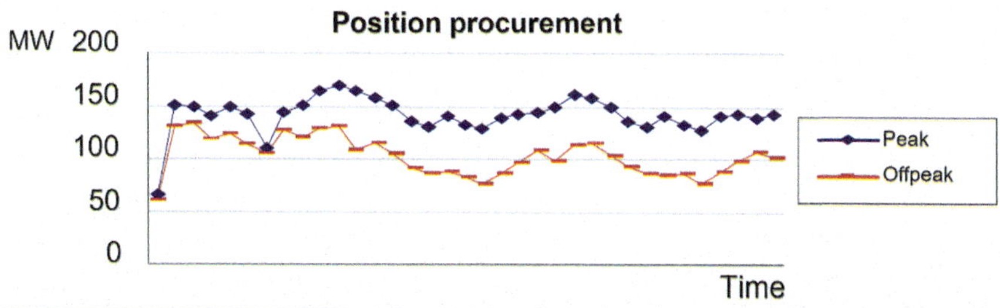

Portfolio own generation	
Value at risk limit	€ 1,500,000
Value at risk (2 days)	€ 1,275,000
Realized Profit since 01/01/y	€ 275,000
Current value of the non-realized position	€ 4,500,000

Fig. 4.20 Example of a risk report

Figure 4.20 shows an example of an extract from a risk management report. It shows the Value-at-Risk, the realised profit, the current value of the non-realised position and other indicators. This information should be presented in an understandable way and with visual appeal. There are professional tools to create such reports, and if they are taken seriously, they can provide vital information that can determine the success or failure of a company.

A risk committee is another instrument of risk management that is intended to ensure the implementation of the procurement strategy and risk policy in the areas of generation, portfolio management and sales, including financial and corporate planning. This committee usually meets roughly every quarter (or less frequently in the case of moderate-sized companies) and is comprised of high-level participants, e.g. the executive board members, managers and some external experts. Among their tasks are:

- Approval of the risk manual.
- Review/outlook risk indicators.
- Suitability of risk management system.
- Adaption of the risk management system.
- Approval of new markets, products, trading partners and trading contracts.
- Approval of current limits.

To sum up, Fig. 4.21 shows the core tools and methods of energy trading and risk management and their role in risk management and control. First, there risk capital that needs to be allocated. Each department/desk/trader is assigned a certain share of the company's total risk capital. This results in a limit system. Monitoring these limits assures that the departments/desk/traders do not exceed their allocated risk capital.

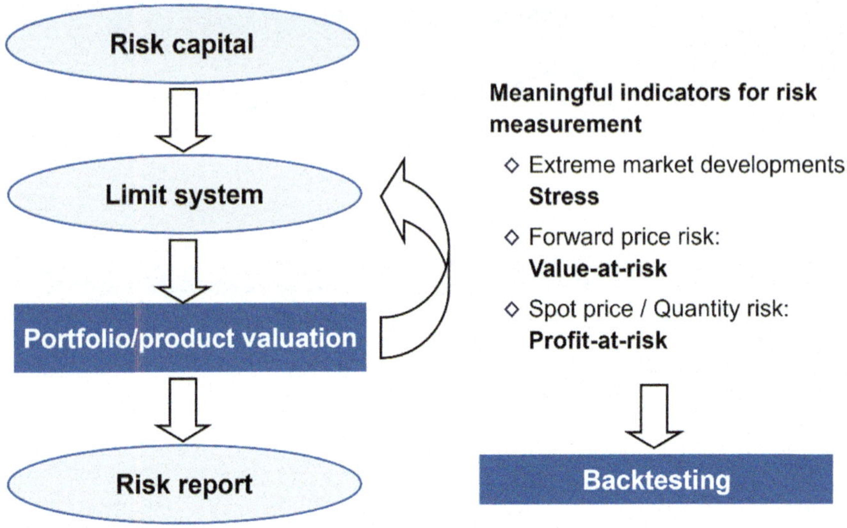

Fig. 4.21 Tools/methods and their role in risk management and control

These results are presented in a risk report that contains the above-mentioned risk indicators. Note however, that these risk indicators can never be purely objective: they are always forward-looking, and the results depend on the chosen methodology, the parameters used, etc. The same applies to the credit assessment, which is not only forward-looking but can also be subjective. The same applies to the credit rating, which is both forward-looking and subjective (which methodology has been chosen, which threshold between groups has been implemented, etc.). Hence, it is recommended to check occasionally whether the chosen assumptions are reasonable. This is referred to as backtesting. Take for instance the calculation of a VaR. A 99% one-day VaR of €X means that with the probability of 99% the portfolio loss after one day will not exceed €X. However, this also means that with a probability of 1% it will exceed €X. In other words: If the VaR has been calculated correctly, the daily loss should exceed the VaR on 1 out of 100 trading days. This (and other risk indicators) can be checked retrospectively via backtesting. The results of these tests should be used to improve methods and models.

Table 4.3 summarises the different tasks of risk management and their frequency of execution.

Table 4.3 Risk management activities and their frequency of execution

Task/activity	Possible frequency
Monitoring of compliance with price risk limits and related reporting including recommendations for corrective measures	At least weekly, depending on the risk attitude or the type and size of the possible open position (for trading companies at least daily)
Credit rating of business partners	At least once a year
Monitoring of compliance with credit limits and related reporting including recommendations for corrective measures	Weekly
Conducting and participating in risk committee meetings	Once a quarter
Monitoring, maintenance, quality assurance and further development of the risk management system	Continuously
Worst-case assessments and determination of risk capital requirements for generation, transport and storage risk	Once a quarter
Approval of markets and products	When it arises
Regular product liquidity checks	Once a quarter
Risk audits for compliance with risk guidelines and the functionality of systems	Once a year
Backtesting of the methods used and revision of guidelines	Once a year

References

Bade, A., Frahm, G., & Jaekel, U. (2009). *A general approach to Bayesian portfolio optimization.* Mathematical Methods of Operations Research, *70*, 337–356. https://doi.org/10.1007/s00186-008-0271-4

Barbu, A., & Zhu, S.-C. (2020). *Monte Carlo methods.* Springer Singapore. https://doi.org/10.1007/978-981-13-2971-5

Börsennews. (2023). *Lexikon - Paketabschlag.* https://www.boersennews.de/lexikon/begriff/paketabschlag/829/

Börse Frankfurt. (2023). *Borussia Dortmund GmbH & Co. KGaA.* Accessed on December 27, 2023. https://www.boerse-frankfurt.de/equity/borussia-dortmund-gmbh-co-kgaa

EFET. (2020). *EECS certificate & National scheme certificate master agreement.* https://data.efet.org/Files/IBOR%20transition/IBOR%20transition%20-%20Outdated%20versions/EFET%20EECS%20MA%20version%201.1%20September%202020%20Final%20-%20full.pdf

Glossary

Term	Page	Short explanation
ACER	117	European Union Agency for the Cooperation of Energy Regulators
Administrative risk	173	The risk of incorrect information or making incorrect decisions or performing incorrect activities because of human error or non-compliance with processes and rules
aFRR	116	Automatic frequency restoration reserve
American option	78	An option that can be exercised at any time before or at the expiration date
Arbitrage	6, 33, 158	Risk-free profits from price differences of the same product at the same time on different marketplaces
Balancing groups	37	A virtual energy account that is managed by the balancing group manager. It can consist of a single or a collective of market participants
Base load	13, 16, 31	Electricity delivery is 24 h on each day of the contract's delivery period
Bid-ask spread	20, 21, 30	The bid-ask spread is defined as the amount by which the lowest ask price exceeds the highest bid price for a commodity
Black Scholes model	140	A model for valuing options with analytical expressions for prices
Broker platform	28	A platform (typically a web application) that enables the trading of forwards and other derivatives
Call option	78	The right to obtain a certain underlying asset at a predetermined price within a certain period (American options) or at a certain point in time (European options)
Carbon emission factor	73	A coefficient telling how much CO_2 is emitted by burning 1 MWh of fuel
CBMP	121	Cross-border marginal price

(continued)

F. Müsgens and A. Bade, *Energy Trading and Risk Management*,
https://doi.org/10.1007/978-3-031-57238-8

(continued)

Term	Page	Short explanation
CHP	85, 101	Combined heat and power
Clean Brown spread	67	A product on the energy markets whose payout results from the difference between the electricity price and the cost of lignite and CO_2 emission certificates
Clean Dark spread	67	A product on the energy markets whose payout results from the difference between the electricity price and the cost of hard coal and CO_2 emission certificates
Clean Spark Spread	67	A product on the energy markets whose payout results from the difference between the electricity price and the cost of natural gas and CO_2 emission certificates
CMOL	121	Common European merit order list
Combined cycle gas turbine (CCGT)	84	A power generation technology that produces electricity by using both gas and steam turbines in a sequential process
Credit exposure	186	The total credit risk is the sum of its two components resale/rebuy risk and free delivery exposure
Credit limit	192	The maximum allowed outstanding expected credit risk with a counterparty
Credit rating	192, 201	The assessment of the creditworthiness of a counterparty
Credit risk	185	The financial loss resulting from the bankruptcy of a counterparty
Current portfolio value	177, 178	The value of the open position at current market prices
Day ahead market	12	The market on which hourly electrical energy products are traded for the next day via an auction
Delivery period	11	The period in which electricity or natural gas is being delivered. It defines the product
EFET	25, 186	European Federation of Energy Traders
ENTSO-E	108, 117, 120	The European Network of Transmission System Operators for Electricity is a European association in which all transmission system operators are compulsory members
EURIBOR	146	The average interest rate at which many European banks lend each other euros
European option	78	An option that can be exercised only at the expiration date
European power exchange (EPEX)	12	An exchange for short-term wholesale electricity trading in several European countries
Exchange	27	A centralized platform or marketplace where financial instruments, commodities, or other assets are bought and sold through organized and regulated trading mechanisms
Extrinsic value	139	The time value of an option resulting from the term of the option and the volatility of the underlying
FCR	93, 105	Frequency Containment Reserve

(continued)

(continued)

Term	Page	Short explanation
Forwards	25, 44, 70	An OTC-traded contract specifying the delivery of an underlying (for instance electrical energy) at a future time or in a future period
Free delivery exposure	186	The value of the energy quantities already delivered but not yet paid for
Front year	156	The calendar year following the current year
Future	23, 24, 25, 133, 194	An exchange-traded contract specifying the delivery of an underlying (for instance electrical energy) at a future time or in a future period
Heavy-tailed	182	A probability distribution is heavy-tailed if it has a higher probability of extreme or rare events, indicating a higher likelihood of observing values that are far from the average
Hedging	33, 145	reducing the risk of an open position by entering a counter position
Historical simulation	177	A method for calculating the Value-at-Risk
Hourly price forward curve (HPFC)	157	The expected future electricity prices for each hour over a specified time horizon
Household customer	40	Customers in the retail market are characterised by low energy volumes and consequently a low absolute revenue per customer
Industrial customer	41	Customers in the retail market that have relatively high energy consumption and generate high absolute revenues per customer
Interest rate swap	146	A swap in which future cash flows are used to amortise a debt (or an investment) from a fixed to a variable interest rate or vice versa
Intrinsic value	139	The current market price of the underlying minus the strike price of an option
In-the-money	79	An option is out-of-the-money when the current market price of the underlying asset is favourable for the option holder to exercise the option
Legal risk	172	The risk that a contract or group of contracts does not contain the desired legal content
Liquidity	30	Question of how quickly and at what cost one can open and close positions in a specific product
Liquidity shortage risk	174	The risk that liquidity will not suffice to meet payment obligations
Long	23, 26, 32, 34, 82	Being long on a product means having bought or owning that product with the expectation that its price will rise over time
Mark-to-Market	156	Valuing something at current market prices

(continued)

(continued)

Term	Page	Short explanation
mFRR	94, 114, 116	Manual frequency restoration reserve
Monte Carlo simulation	145, 177, 179	A computational technique that uses random sampling to model and analyse complex systems or processes, providing a range of possible outcomes and their probabilities
Normal distribution	179	A continuous probability distribution
Open cycle gas turbine (OCGT)	99	A type of gas turbine power plant that generates electricity by burning natural gas in an open cycle process
Open position	36, 156, 199	The difference between what has been bought and what has been sold for a certain delivery period h
Option	77	gives someone the right, but not the obligation, to buy or sell something at a predetermined price at a specific time (or time points) in the future
Order book	21	This contains all bids and asks for a certain product
Out-of-the-money	79, 139	An option is out-of-the-money when the current market price of the underlying asset is unfavourable for the option holder to exercise the option
Over-the-counter (OCT)	25	Bilateral trading where two entities agree on terms of a business deal. This deal may have been arranged by a broker
Peak load	13, 31	Electricity delivery from 8:00 to 20:00 on each weekday of the contract's delivery period
Price risk	170, 174, 194, 201	The risks of future price variations changing the value of the open position
Product liquidity risk	172, 194	The risk that the market for a product becomes illiquid
Proprietary trading (prop trading)	2	The practice where a company trades financial instruments, such as power futures or derivatives, using its own funds to generate profits for the firm
Put option	78, 83	The right to sell a certain underlying asset at a predetermined price within a certain period (American options) or at a certain point in time (European options)
Real optionality	85	The option value of a power plant arising from its flexibility
Resale/rebuy risk	186	The difference between the current market price and the contracted price of undelivered energy
Risk capital	174, 175	The amount of capital that a company can afford to lose and that is assigned to different risky activities within the company
Risk indicator	175	A measurable and observable factor or variable that provides insights into the likelihood or potential impact of a specific risk

(continued)

(continued)

Term	Page	Short explanation
Risk management process	198	The process of managing risk properly
Risk-free interest rate	141	The theoretical rate of return on an investment with no risk of financial loss
RR	119, 124	RR is the abbreviation for replacement reserve. This is also known as quaternary reserve and is a reserve in the power grid that helps to keep the grid frequency stable
Schedule	27, 54	An important class of non-standardised products, defined as an energy time series
Short	26, 32, 34	Being short on a product means having sold or borrowed that product with the expectation that its price will rise over time
Spark spread	66	Non-standardised, cross-commodity product on the energy markets. Its payout is the difference between the electricity price and a fuel price
Speculation	33	Strategy where a position was opened in anticipation of a certain market price development
Spread	66	The difference between two prices
Standard deviation	141, 180, 181	Second moment of a random variable can be a measure of volatility
Stress value	176	The change in the value of a portfolio from one day to another day in an expected worst case
Strike price	78, 79, 140	The price at which the underlying an option can be bought (call option) or sold (put option)
Student's-*t*-distribution	182	A continuous probability distribution
Swap	70	An agreement to exchange cash flows at specified future times according to a prearranged formula
System risk	173	The risk of malfunctions within the processes, systems and procedures of the company
Trading period	10	The period during which an electricity or natural gas product can be being traded
Transmission system operator (TSO)	37, 93, 106, 108	A regulated entity responsible for the reliable and efficient operation of the electrical grid
Value-at-Risk (VaR)	47, 176, 178, 203	A statistical measure representing the maximum potential loss, within a specified confidence level, that a portfolio may experience over a given time horizon
Volume risk	172	The financial risk arising from a deviation between expected and realised supply and/or demand volumes

The manufacturer's authorised representative in the EU is Springer Nature Customer Service Centre GmbH, Europaplatz 3, 69115 Heidelberg, Germany. If you have any concerns regarding our products, please contact ProductSafety@springernature.com

Printed and bound by CPI Group (UK) Ltd, Croydon, CR0 4YY
30/09/2026
02225586-0002